EXPANDED & UPDATED

MATHEMATICAL
Quickies &
Trickies

Over 300 Nonroutine Questions
to Enhance Your Problem-Solving Skills

Yan Kow Cheong

MATHPLUS
PUBLISHING
Specialists in Mathematics Education

MATHPLUS Publishing
Blk 639 Woodlands Ring Road
#02-35 Singapore 730639

E-mail: publisher@mathpluspublishing.com
Website: www.mathpluspublishing.com

First published in the United States in 2018

National Library Board, Singapore Cataloguing-in-Publication Data

Yan, Kow Cheong.
    Mathematical quickies & trickies / Yan Kow Cheong. – Expanded & updated ed. – Singapore : MathPlus Pub., 2011.
    p. cm.
    ISBN : 978-981-08-5412-6 (pbk.)

    1. Mathematical recreations. I. Title.

QA95
793.74 -- dc22   OCN668054034

Printed in the United States of America

A **mathematical quickie** is a problem which may be solved by laborious methods, but which with proper insight may be disposed of quickly. This term was coined by the late Professor Charles W. Trigg to describe problems that yield almost instantly to a flash of inspiration.

A **mathematical trickie** is a problem whose solution rests on some key word, phrase, or idea rather than on a mathematical routine. Most number riddles would thus qualify as trickies.

Although there exist thousands of these quickies and trickies in the recreational mathematics literature, many of them require a fairly sophisticated level of mathematics from the problem solver.

To expose both upper primary and lower secondary school (grades 5–8) students to some of these entertaining nonroutine questions, and to arouse their interests, over 300 elementary quickies and trickies are compiled from the fields of arithmetic, geometry, algebra, and recreational mathematics.

Ranging from the simple and trivial to the complex and challenging, most of these problems and solutions should prove accessible to the average primary (or elementary) school student. However, some of these trick and tricky problems may pose a challenge even to the talented or gifted secondary student.

Sources for most of the problems are difficult to trace. However, I have provided a reference list from which most of the problems were taken, all being altered in one form or another to render the language contemporary.

The challenge of **Mathematical Quickies & Trickies** is not only to solve these mathematical brainteasers, but also to come up with more elegant solutions than the ones provided.

I would be glad to hear from readers who would like to share with me their ingenious solutions. You may contact me at kcyan.mathplus@gmail.com and visit my blogs at http://singaporemathplus.com and http://singaporemathplus.net or follow me on Twitter as @MathPlus and @Zero_Math.

K C Yan

CONTENTS

	Preface	iii
1	Mental Computation 1	1
2	Mental Computation 2	10
3	Number Series	17
4	A Tricky Way with Fractions	24
5	Test Your Calculator Proficiency	31
6	Simplifying a Complex Fraction	39
7	Recurring (Repeating) Decimals	45
8	Is Zero an Even or Odd Integer?	51
9	Casting Out Nines	57
10	Be a Calculator Expert	64
11	Division by 9	71
12	Mathematical Trickies (Number Riddles)	77
13	Ten Steps to be Math Smart	83
14	Shortcuts à la *Trachtenberg*	88
15	Geometrical Quickies 1	94
16	Geometrical Quickies 2	99
17	Geometrical Quickies 3	104
18	(Sugar + Coffee) + Milk = Sugar + (Coffee + Milk)	110
19	More Applications of Number Laws	117
20	Law of One	123
21	Distributive Law	129
22	More Applications of Distributive Law	136
23	The Joy of Guesstimation	142
24	Are You a Fermi Disciple?	147
25	Bravo Singapore	154
26	Lightning Calculators	160
27	Geometrical Quickies 4	167
28	Some Calculator Quickies	174
	Answers	181
	Bibliography & References	201

1 Mathematical Quickies & Trickies

Mental Computation 1

Example 1

Mentally compute the following.

(a) $473 + 99 = \boxed{}$

Method 1

$$473 + 9 = 473 + (100 - 1)$$
$$= (473 + 100) - 1$$
$$= 573 - 1$$
$$= 572$$

Method 2

$$473 + 99 = 472 + 1 + 99$$
$$= 472 + 100$$
$$= 572$$

(b) $976 + 1005 = \boxed{}$

Method 1

$$976 + 1005 = 976 + (1000 + 5)$$
$$= (976 + 1000) + 5$$
$$= 1976 + 5$$
$$= 1981$$

Look for "friendly numbers" like 10, 100, 1000, ... — those powers of 10.

Method 2

$$976 + 1005 = 976 + 5 + 1000$$
$$= 981 + 1000$$
$$= 1981$$

(c) $1234 + 9995 = \boxed{}$

Method 1

$$1234 + 9995 = 1234 + (10,000 - 5)$$
$$= 1234 - 5 + 10\ 000$$
$$= 1229 + 10,000$$
$$= 11\ 229$$

Method 2

$$1234 + 9995 = 1229 + 5 + 9995$$
$$= 1229 + 10,000$$
$$= 11,229$$

(d) $993 + 994 = \boxed{}$

Method 1

$$993 + 994 = (1000 - 7) + (1000 - 6)$$
$$= 1000 + 1000 - 13$$
$$= 2000 - 13$$
$$= 1987$$

Method 2

$$993 + 994 = 993 + 7 + 994 + 6 - 7 - 6$$
$$= (1000 + 1000) - (7 + 6)$$
$$= 2000 - 13$$
$$= 1987$$

(e) $9992 + 9999 = \boxed{}$

Method 1

$$9992 + 9999 = 9991 + 1 + 9999$$
$$= 9991 + 10,000$$
$$= 19,991$$

Method 2

$$9992 + 9999 = 9992 + 10,000 - 1$$
$$= 19,992 - 1$$
$$= 19,991$$

Method 3

$$9992 + 9999 = (10,000 - 8) + (10,000 - 1)$$
$$= (10,000 + 10,000) - 8 - 1$$
$$= 20,000 - 9$$
$$= 19,991$$

Example 2

Mentally compute the following.

Beware of the change in sign!

(a) $995 - 97 = \boxed{}$

Method 1

$$995 - 97 = (1000 - 5) - (100 - 3)$$
$$= (1000 - 100) - 5 + 3$$
$$= 900 - 2$$
$$= 898$$

Method 2

$$995 - 97 = 995 - 100 + 3$$
$$= 895 + 3$$
$$= 898$$

(b) 9994 – 997 = ☐

Understand why it is –6 + 3, and not –6 – 3.

Method 1

$$9994 - 997 = (10{,}000 - 6) - (1000 - 3)$$
$$= (10{,}000 - 1000) - 6 + 3$$
$$= 9000 - 3$$
$$= 8997$$

Method 2

$$9994 - 997 = 9994 - 1000 + 3$$
$$= 8994 + 3$$
$$= 8997$$

Method 3

$$9994 - 997 = 9994 - 994 - 3$$
$$= 9000 - 3$$
$$= 8997$$

(c) 980 – 93 = ☐

Method 1

$$980 - 93 = 980 - 100 + 7$$
$$= 880 + 7$$
$$= 887$$

Method 2

$$980 - 93 = 980 - 80 - 13$$
$$= 900 - 13$$
$$= 887$$

Method 3

$$980 - 93 = (1000 - 20) - (100 - 7)$$
$$= (1000 - 100) - 20 + 7$$
$$= 900 - 13$$
$$= 887$$

(d) 199,996 – 99,998 = ☐

Method 1

$$199{,}996 - 99{,}998 = 199{,}996 - 100{,}000 + 2$$
$$= 99{,}996 + 2$$
$$= 99{,}998$$

Beware of the change in sign.

Method 2

$$199{,}996 - 99{,}998 = 199{,}996 - 99{,}996 - 2$$
$$= 100{,}000 - 2$$
$$= 99{,}998$$

Method 3

$$199{,}996 - 99{,}998 = (200{,}000 - 4) - (100{,}000 - 2)$$
$$= (200{,}000 - 100{,}000) - 4 + 2$$
$$= 100{,}000 - 2$$
$$= 99{,}998$$

(e) $699{,}992 - 99{,}988 =$ ☐

Method 1

$$699{,}992 - 99{,}988 = 699{,}992 - 100{,}000 + 12$$
$$= 599{,}992 + 12$$
$$= 599{,}992 + 8 + 4$$
$$= 600{,}000 + 4$$
$$= 600{,}004$$

Method 2

$$699{,}992 - 99{,}988 = 699{,}988 + 4 - 99{,}988$$
$$= 699{,}988 - 99{,}988 + 4$$
$$= 600{,}000 + 4$$
$$= 600{,}004$$

Method 3

$$699{,}992 - 99{,}988 = (700{,}000 - 8) - (100{,}000 - 12)$$
$$= (700{,}000 - 100{,}000) - 8 + 12$$
$$= 600{,}000 + 4$$
$$= 600{,}004$$

PRACTICE

1. Mentally do the following.
    (a) $578 + 98$ (b) $896 + 1002$
    (c) $6543 + 9997$ (d) $9994 + 9998$

2. Mentally do the following.
    (a) $897 - 97$ (b) $9997 - 998$
    (c) $986 - 97$ (d) $199{,}995 - 99{,}997$

3. Evaluate the following.
    (a) $9876 - 997 + 98$ (b) $100{,}002 + 997 - 9996$

1. A *Math Clinic* has all its employees working an equal number of hours each day. The total number of working hours for a day is 133. How many employees are there?

2. The total age of two sisters, Esther and Ruth, is 14. Esther is 13 years older than Ruth. How old will Ruth be in 12 years' time?

3 Two trains leave at the same time from two cities 425 km apart towards each other at 55 km/h and 75 km/h. What will be the distance between the two trains half an hour before they meet each other?

4 Solomon and David played poker between them and the stake for each game is $1. At the end of their games, Solomon won $3 and David won 3 games. How many games were played?

Being a professional poker player is perfectly legal!

5 A bottle of lemonade costs 40 cents and the lemonade costs 30 cents more than the bottle. How much does the empty bottle and the lemonade each cost?

6 Two mothers and two daughters shared $300. Each received $100. How is this possible?

No stepmother or godmother is involved!

7 Train ACE leaves Samaria for Antioch at 8:00 a.m., traveling at 120 km/h. Another train XUZ leaves Antioch for Samaria at 8:00 a.m., traveling at 150 km/h. The distance between the two towns is 1000 km. When the two trains meet each other, which train is nearer to Samaria?

8 Mr. and Mrs. Pilate had two daughters and each of them had two children. How many members are there in the family altogether?

9 An unscrupulous seller has his meter ruler used for measuring cloth 4 cm short. This means for every one meter of cloth you buy, you actual get 96 cm of it.

What will be the actual length of cloth if you buy $3\frac{1}{2}$ meters of cloth?

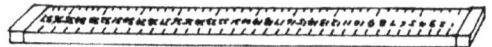

A tampered meter ruler

After his last Mathematics paper which had forced him to keep awake 2 days, Joseph decided to go to bed at 9:00 p.m., and set the alarm to wake him up at 11:00 a.m. the next day. How many hours of sleep did he get?

A Prime Dream

2 Mathematical Quickies & Trickies

Mental Computation 2

Example 1

Mentally compute the following.

(a) $123 + 298 = \boxed{}$

Method 1

$$123 + 298 = 123 + (300 - 2)$$
$$= (123 + 300) - 2$$
$$= 423 - 2$$
$$= 421$$

Method 2

$$123 + 298 = 121 + 2 + 298$$
$$= 121 + 300$$
$$= 421$$

(b) $234 + 896 = \boxed{}$

Method 1

$$234 + 896 = 234 + (900 - 4)$$
$$= (234 + 900) - 4$$
$$= 1134 - 4$$
$$= 1130$$

Method 2

$$234 + 896 = 230 + 4 + 896$$
$$= 230 + 900$$
$$= 1130$$

(c) $798 + 456 = \boxed{}$

Method 1

$$798 + 456 = 798 + 2 + 454$$
$$= 800 + 454$$
$$= 1254$$

Method 2

$$798 + 456 = (800 - 2) + 456$$
$$= (800 + 456) - 2$$
$$= 1256 - 2$$
$$= 1254$$

(d) 3307 + 4998 = ▢

Method 1

3307 + 4998 = 3307 + (5000 − 2)
$\qquad\qquad$ = (3307 + 5000) − 2
$\qquad\qquad$ = 8307 − 2
$\qquad\qquad$ = 8305

Method 2

3307 + 4998 = 3305 + 2 + 4998
$\qquad\qquad$ = 3305 + 5000
$\qquad\qquad$ = 8305

(e) 6548 + 792 = ▢

Method 1

6548 + 792 = 6548 + (800 − 8)
$\qquad\qquad$ = (6548 + 800) − 8
$\qquad\qquad$ = 7348 − 8
$\qquad\qquad$ = 7340

Method 2

6548 + 792 = (6540 + 8) + 792
$\qquad\qquad$ = 6540 + (8 + 792)
$\qquad\qquad$ = 6540 + 800
$\qquad\qquad$ = 7340

> ## CAUTION
>
> *Mathematical Quickies & Trickies*
> *May Improve Your*
> *Mathematical Health!*

1. A drawer contains 21 black and 19 grey socks. It is pitch dark. How many socks do you have to remove to ensure a matching pair?

How many socks make a pair?

2. How many months have 30 days?

'Green' RECYCLED CALENDARS!

GREAT SAVINGS!

1937

Year 1901   2013   2069 Calendar

③ What is the largest number you can make with just three whole numbers?

It can't be 999! Why not call the police for the answer?

④ A chess club had 64 participants for its annual elimination tournament. In the first round, 32 games were played to eliminate half the contestants. In the second round a further quarter were eliminated, and so on until the winner emerged. How many games were played?

5 You roll a normal dice 9 times and it comes up six each time. What is the chance it will come up six the 10th time?

> Opposite faces of a normal dice sum up to 7.

6 Five apples in a basket are to be divided among five girls so that each girl gets an apple, and yet one apple remains in the basket. How should the division take place?

7 A bike owner with 19 wheels decides to make 7 bicycles and tricycles from them. How many of each type can he make to use up all the wheels?

8 Joe recalled before closing his geometry book that the product of the page numbers was 1260. What were the two open pages?

9   At 6 o'clock the clock strikes 6 times. My watch records the time between the first and last strokes to be 30 seconds. How long will the clock take to strike 12 at noon?

10   The *I Love Mathematics* costs $1 plus half its price.
How much does the popular book cost?

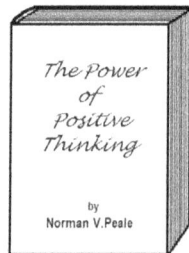

The Power
of
Positive
Thinking

by
Norman V. Peale

10 million copies sold

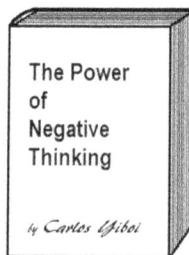

The Power
of
Negative
Thinking

by Carlos Yibei

10,000 copies sold

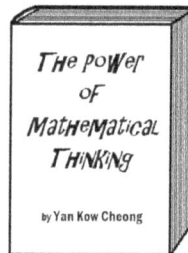

The power
of
Mathematical
Thinking

by Yan Kow Cheong

? copies sold

Mathematical Quickies & Trickies

Number Series

Example 1

Without a calculator, find the value of the following.

(a) $3 + 33 + 333 + 3333 + \cdots + 3333333333$
(b) $1 - 2 + 3 - 4 + 5 - 6 + \cdots + 1995 - 1996 + 1997 - 1998 + 1999$

Solution:

(a) $3 + 33 + 333 + 3333 + \cdots + 3333333333$
$= 3 \times (1 + 11 + 111 + 1111 + \cdots + 1111111111)$
$= 3 \times (123456789 + 1111111111)$
$= 3 \times 1,234,567,900$
$= 3,703,703,700$

(b) $1 - 2 + 3 - 4 + 5 - 6 + \cdots + 1995 - 1996 + 1997 - 1998 + 1999$

Method 1
$1 - 2 + 3 - 4 + 5 - 6 + \cdots + 1995 - 1996 + 1997 - 1998 + 1999$
$= (1 - 2) + (3 - 4) + (5 - 6) + \cdots + (1995 - 1996) + (1997 - 1998) + 1999$
$= (-1) \times (1998 \div 2)$

$= -999 + 1999$
$= 1000$

Method 2
$1 - 2 + 3 - 4 + 5 - 6 + \cdots + 1995 - 1996 + 1997 - 1998 + 1999$
$= 1 + (-2 + 3) + (-4 + 5) + \cdots + (-1996 + 1997) + (-1998 + 1999)$
$= 1 + (1) \times (1998 \div 2)$

$= 1 + 999$
$= 1000$

Method 3

$1 - 2 + 3 - 4 + 5 - 6 + \ldots + 1995 - 1996 + 1997 - 1998 + 1999$

$= (999 + 1001) - 1000$ $(1 + 1999 = 2000; -2 - 1998 = -2000;$

$= 2000 - 1000$ $3 + 1997 = 2000; -4 - 1996\ldots)$

$= 1000$

Example 2

What is the value of

$$\left(1 - \frac{1^2}{100}\right) \times \left(1 - \frac{2^2}{100}\right) \times \left(1 - \frac{3^2}{100}\right) \times \cdots \times \left(1 - \frac{2011^2}{100}\right)?$$

Solution:

Since one of the factors $\left(1 - \frac{10^2}{100}\right)$ is 0, the product is 0.

PRACTICE

What is the value of

(a) $1 + 2 + 3 + \cdots + 2009 + 2010$,

(b) $2 + 4 + 6 + \cdots + 2012$,

(c) $1 + 2 - 3 - 4 + 5 + 6 - 7 - 8 + \cdots - 1999 - 2000$.

1. Which is worth more: a pound of $10 gold pieces
                                or
             half a pound of $20 gold pieces?

2. When my father was 43, I was 13. Now he is twice as old as I am.
   How old am I?

3. A boy has as many sisters as brothers, but each girl has only half as many
   sisters as brothers. How many boys and girls are there in the family?

4 A half is a third of a number.
   What is the number?

5 An alarm clock runs 4 minutes slow every hour. It was set right $3\frac{1}{2}$ hours ago. Now another clock, which is correct, shows noon. In how many minutes, to the nearest minute, will the alarm clock show noon?

6 A donkey travels half his route, with no load, at 12 km/h. He completed the rest of the journey with a load which slowed him to 4 km/h. What is his average speed?

The Wisdom of Donkeys!

* 7 A train moving at 55 km/h meets and is passed by a train moving at 45 km/h. A passenger in the first train sees the second train take 6 seconds to pass her. How long is the second train?

8 After cycling for two thirds of the journey, John got a puncture. He walked the remaining distance taking twice as long as he cycled. How many times as fast did he cycle as walk?

9 Jane and Joe want to buy pens. However, Joe is 22 cents short of the price of a pen, and Jane is 2 cents short. They decide to pool their money and buy a single pen. But again, they still do not have enough money. How much does a pen cost?

* 10 In a storeroom, 1000 kg of fresh grapes are stored. The grapes contained 99% water when fresh, but one week later, because of the dryness of the grapes, there was only 98% water. How much do the grapes weigh now?

Mathematical Quickies & Trickies

A tricky way to add two fractions

Example 1

Add $\frac{a}{b} + \frac{c}{d}$.

$$\frac{a}{b} + \frac{c}{d} = \frac{(a \times d) + (b \times c)}{b \times d}$$

e.g., $\quad \dfrac{2}{5} + \dfrac{3}{7} = \dfrac{(2 \times 7) + (3 \times 5)}{5 \times 7}$

$$= \frac{14 + 15}{35}$$

$$\frac{2}{5} + \frac{3}{7} = \frac{29}{35}$$

Why not $\dfrac{a}{b} + \dfrac{c}{d} = \dfrac{a+c}{b+d}$?

A tricky way to subtract two fractions

Example 1

Subtract $\dfrac{a}{b} - \dfrac{c}{d}$.

$$\frac{a}{b} - \frac{c}{d} = \frac{(a \times d) - (b \times c)}{b \times d}$$

e.g., $\dfrac{7}{9} + \dfrac{3}{5} = \dfrac{(7 \times 5) + (3 \times 9)}{9 \times 5}$

$$= \frac{35 - 27}{45}$$

$$\frac{7}{9} - \frac{3}{5} = \frac{8}{45}$$

When can $\dfrac{a}{b} - \dfrac{c}{d} = \dfrac{a-c}{b-d}$?

1. If one corner is cut off a rectangle, how many corners are left?

2. Utility-hole covers are round and not some polygonal shape.
   Is there a practical reason for this?

Why are man-holes circular?

3 A clock takes 2 seconds to strike 2 o'clock. How long will it take to strike 3 o'clock?

CAUTION

Simple-sounding questions seldom have obvious answers!

4 All my pets are dogs except two, all my pets are cats except two, and all my pets are hamsters except two. How many dogs do I have?

* **5** There are two flasks filled to the same level, one full of oil, the other full of water. Take a spoonful of oil, put it into the water flask and stir. Then take a spoonful of the resultant mixture and put it back in the oil flask. Will there be more oil in the original water flask or more water in the oil flask?

> Imagine the molecules of oil or water to be the sizes of balls!

6 A tent was pitched between 4 trees that formed a square. How would a square tent twice the floor area between the same 4 trees be pitched?

7. A beggar's brother died. The man who died had no brother. Explain how this was possible.

8. A man travels from town A to town B at 30 km/h. How fast does he have to travel from B to A to average 60 km/h for the total journey from A to B and back again?

9. An empty barrel weighs 10 kilograms. What can be put in the barrel to make it weigh 9 kilograms?

10. With a 7-minute sand timer and an 11-minute sand timer, what is the easiest way to time the boiling of an egg for 15 minutes?

5 Mathematical Quickies & Trickies

Test Your Calculator Proficiency

Example 1

Using only the $\boxed{1}$, $\boxed{+}$ and $\boxed{=}$ keys, how many presses do you need to get the number 12 in your calculator display?

Solution

The challenge is to press as few keys as possible to form the number 12.
To do this, these keys may be pressed:

$$\boxed{1}\ \boxed{1}\ \boxed{+}\ \boxed{1}\ \boxed{=}$$

5 presses are needed to get 12.

Example 2

What is the sum of all the numbers from 1 to 50, i.e., $1 + 2 + 3 + \cdots + 49 + 50$?

Solution

Method 1

A quicker way is to pair up the numbers between 1 and 50:

$$
\left.
\begin{array}{l}
1 + 50 \ = 51 \\
2 + 49 \ = 51 \\
3 + 48 \ = 51 \\
\ldots \\
25 + 26 = 51
\end{array}
\right\}
\quad
\begin{array}{l}
25 \text{ pairs of } 51\text{s} \\
= 25 \times 51 \\
= 1275
\end{array}
$$

Therefore, $1 + 2 + 3 + \cdots + 49 + 50 = 1275$

What about a visual proof?

Method 2

Imagine a square made up of $1 + 2 + 3 + \cdots + 49 + 50 + 49 + 48 + \cdots + 3 + 2 + 1$ dots.

Number of dots in the square $= 2 \times (1 + 2 + 3 + \cdots + 48 + 49) + 50$

$$= \text{square of } 50 \times 50 \text{ dots}$$
$$= 2500$$

Now, $2 \times (1 + 2 + 3 + \cdots + 48 + 49) = 2500 - 50$
$$= 2450$$

$(1 + 2 + 3 + \ldots + 48 + 49) = \dfrac{2450}{2} = 1225$

Hence, $1 + 2 + 3 + \cdots + 48 + 49 + 50 = 1225 + 50 = 1275$

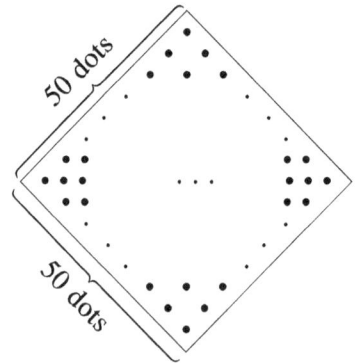

50 dots

50 dots

Can you find the same sum using two other methods of solutions?

Example 3

Using only these keys, how would make the calculator display 100?

| 3 | 7 | + | – | = |

What is the least number of presses you would need?

Solution

| 3 | 7 | + | 7 | 3 | – | 7 | – | 3 | = |

10 presses are needed to get the number 100.

$$37 + 73 - 7 - 3$$
$$= 110 - 10$$
$$= 100$$

Alternatively, the following keys may also be pressed:

| 7 | 7 | + | 3 | 3 | – | 7 | – | 3 | = |

Are there other possible key presses to yield 100?

1 On keying a four-digit number into a calculator, Moses unintentionally added in a "1" in front of the number. What number must he subtract so that he would end with the required number?

2 Using any combination of +, ×, −, ÷ and all of 1, 3, 3 and 5 once only, make the number 36.

3 Ann accidentally enters 89 × 7348 instead of 89 × 7349. How much should she add or subtract in order to get the correct answer?

④ Using the digits 1, 2, 3, 4, 5, and 6, what three two-digit numbers which when multiplied together give the smallest product?

⑤ What number comes next in this series?

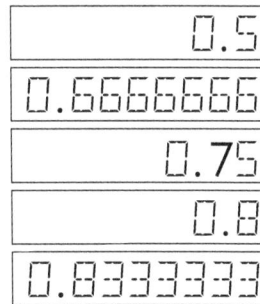

0.5
0.6666666
0.75
0.8
0.8333333

6   When Lisa taps the following keys

| 1 | 2 | + | 2 | 5 | % | = |

into her calculator it shows

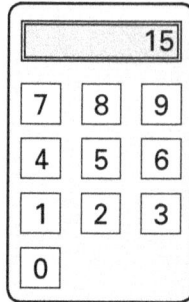

15
7	8	9
4	5	6
1	2	3
0

She expects this because:

$$12 + (\text{one quarter of } 12) = 12 + 3 = 15$$

However, her father's calculator (which he uses for work every day) gives the answer 16. The calculator is functioning correctly and the same calculator is widely available for sale. What is her father's profession?

7   Paul accidentally keys in 28 × 9978 into his four-function calculator instead of 28 × 9998. Does he need to add or subtract to arrive at the correct answer?

8   When the numbers 807540 are keyed in, the numbers 5204182710 appear on the display of a broken calculator. Using the same calculator, what will the numbers 10654 be converted to?

| 5204182710 |
7	8	9
4	5	6
1	2	3
0

9 Where should one insert the number four that is currently missing from the following sequence?

$$172350608$$

10 Enter 38 on a calculator. Now reduce it to 1 by using any operation key and the number 3 only. What is the fewest number of key presses needed?

11 Look at the following:

$$2 + 47 = 49$$
$$2 \times 47 = 94$$

What is so special about these multiplications?
Can you find other pairs that exhibit the same property?

6 Mathematical Quickies & Trickies

Simplifying a Complex Fraction

Given the complex fraction $\dfrac{\frac{a}{b} + \frac{c}{d}}{\frac{u}{v} + \frac{x}{y}}$, the average student is often seen simplifying it as follows:

$$\frac{\frac{a}{b} + \frac{c}{d}}{\frac{u}{v} + \frac{x}{y}} = \frac{\frac{ad + bc}{bd}}{\frac{uy + vx}{vy}} = \frac{ad + bc}{bd} \div \frac{uy + vx}{vy}$$

$$= \frac{ad + bc}{bd} \times \frac{vy}{uy + vx}$$

$$= \frac{advy + bcvy}{bduy + bdvx}$$

An Effective Way

A shorter way to simplify the complex fraction is as follows:

Multiplying both the numerator and denominator of the complex fraction by the LCM of the simple fractions provides an efficient method of simplification.

$$\frac{\frac{a}{b} + \frac{c}{d}}{\frac{u}{v} + \frac{x}{y}} = \frac{\left(\frac{a}{b} + \frac{c}{d}\right)bdvy}{\left(\frac{u}{v} + \frac{x}{y}\right)bdvy} = \frac{advy + bcvy}{bduy + bdvx}$$

1. How many rectangles are there in this figure?

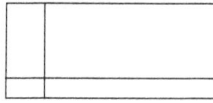

2. Mentally compute the following:
   (a) 1235 + 9996,
   (b) 9997 − 998.

3. Ah Koon has reached the age of 21 after passing only five birthdays. How can he be so much more than five years on his fifth birthday?

④ Fourteen pipes of lengths 6 m and 7 m were laid over a distance of 93 m. How many pipes of each kind were laid?

⑤ Mr. Kiasu had to give out 8 more notes if he were to pay his Algebra book with 3-dollar notes instead of 5-dollar notes. How much does the book cost?

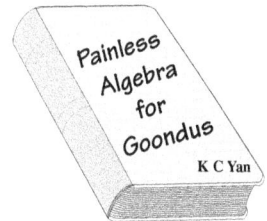

Painless Algebra for Goondus

K C Yan

6. A yellow bulb flashes every 2 minutes and a blue bulb flashes every $3\frac{1}{2}$ minutes. If both bulbs start flashing together at 9.00 a.m., what time after 10.00 a.m. will both bulbs flash together again?

7. There are 80 marbles in two boxes. If 27 marbles from the first box are put into the second box, each box will have the same number of marbles. How many more marbles has the first box than the second box?

8   A train travels at 38 km/h, and a car travels at 57 km/h. The train left town *A* seven hours before the car, but the car overtook the train and arrived at town *B* two hours before the train. Find the distance from *A* to *B*.

9   Before the end of the day there remains $\frac{4}{5}$ of what has elapsed since the day began. What time is it now?

I don't have time! I need a 25-hour day! Yes, I need a Christmas day!

10 A cyclist traveled from town P to town Q at 20 km/h, and went back at 10 km/h. What is her average speed for the whole journey?

Mathematical Quickies & Trickies

7

Recurring (Repeating) Decimals

Example 1

What is the 100th digit in the decimal for $\frac{1}{7}$?

Solution:

$$\frac{1}{7} = 0.142857\ 142857\ 142857\ldots$$

$$= 0.\overline{142\ 857} \quad \text{(repeats in blocks of 6 digits)}$$

$$100 = 6 \times 16 + 4$$

The first 100 digits will include 16 blocks of 6 digits plus the first four digits of the block.

$$\frac{1}{7} = \underbrace{0.142857\ 142857\ \ldots}_{\text{16 blocks}} \qquad \underset{\underset{\text{100th digit}}{\uparrow}}{1428}$$

The 100th digit in the decimal for $\frac{1}{7}$ is 8.

Example 2

How many digits precede the one hundredth 7 in the decimal for $\frac{1}{7}$?

Solution:

The digit 7 appears at the end of every block of 6 digits.

$$\frac{1}{7} = 0.142857\ 142857\ 142857\ \ldots\ 142857\ 142857$$

$$\underset{\text{1st 7}}{\uparrow}\quad \underset{\text{2nd 7}}{\uparrow}\quad \underset{\text{3rd 7}}{\uparrow}\qquad \underset{\text{99th 7}}{\uparrow}\quad \underset{\text{100th 7}}{\uparrow}$$

Number of digits preceding the 100th 7 = $(6 \times 100) - 1$ or $6 \times 99 + 5$

$$\begin{aligned} &= 600 - 1 & &= 594 + 5 \\ &= 599 & &= 599 \end{aligned}$$

1. There are 4 volumes of *Mathematical Goodies* on a library shelf in proper order, with Volume 1 on the left. Each volume is 2 cm thick. A bookworm travels from the front cover of Volume 1 and eats its way straight through to the back cover of Volume 4. How far did it travel?

2. How could two men play five games of chess and each win the same number of games without any ties?

3 A cake is to be cut into eight equal pieces. What is the least number of knife strokes required?

4 What is the radius of the circle?

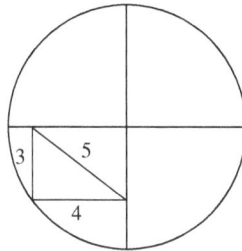

⑤ In which order are the numbers: 0, 2, 3, 6, 7, 1, 9, 4, 5, 8 arranged?

* ⑥ Jack needs 5 cigarette butts to make a standard budget cigarette. How many cigarettes can he assemble and smoke if he finds 25 butts?

7   There are some marbles and some boxes. If one marble is put in each box, one marble will be left without a box. If two marbles are put in each box, one box will remain empty. How many marbles and how many boxes are there?

8   Farmer Zin's livestock contains chickens and rabbits. Together they have 35 heads and 94 feet. How many chickens and how many rabbits are there?

9   It is 9 p.m. now. What time will it be 23,999,999,996 hours later?

10   An angle measures $2\frac{1}{2}^{\circ}$. What percentage increase will it look through a glass of magnifying power three times?

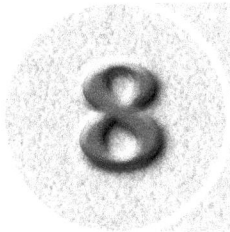

Mathematical Quickies & Trickies

8

Is Zero an Even or Odd Integer?

Misconception: Zero is neither even nor odd.

That "zero is even" is not an arbitrary decision may be argued as follows:

1. Even numbers are defined as those numbers that are divisible by 2; odd numbers leave a remainder of 1 when divided by 2. On this definition, zero must be even.

2. The sum of any two even numbers is even.
   The sum of an even number and an odd number is odd.
   The sum of two odd numbers is even.

   None of these generalizations would be correct if zero were not unambiguously even.

3. The product of any two even numbers is even.
   The product of an even number and an odd number is even.
   The product of two odd numbers is odd.

   None of these generalizations would be correct if zero were not unambiguously even.

   Some common properties of zero are:

   1. 0 is *not* a counting number.

   2. 0 *is* a whole number.

   3. $\frac{1}{n} = 0$, $n \neq 0$.

   4. $\frac{n}{0}$ is undefined, $n \neq 0$.

   5. $\frac{0}{0}$ is indeterminate.

Looks like zero can be anything but nothing!

1　What is the smallest number of cuts needed to cut a 10 cm by 10 cm by 10 cm block of wood into 1-cm cubes?

2　If 5 prisoners are locked in a cell, 4 will be left without a bed.
If 6 prisoners are locked in a cell, 2 beds will be unoccupied.
How many prisoners and how many cells are there?

Numeracy in captivity can only trigger creativity!

Witness: I left a $50 note between pages 145 and 146 of the book and the defendant stole it.

Judge: Case dismissed.

Can you explain why?

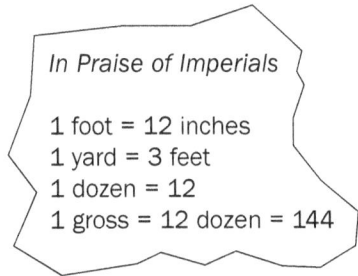

> *In Praise of Imperials*
>
> 1 foot = 12 inches
> 1 yard = 3 feet
> 1 dozen = 12
> 1 gross = 12 dozen = 144

④ A frog is at the bottom of a 30-foot well. Each hour it climbs 3 feet and slips back 2 feet. How long would it take for the frog to get out?

⑤ Dr. Pong fenced his 248 chickens in a square-shaped enclosure. If he needed 33 fence poles on each side of the square, how many poles were used?

6 What is the angle between the hour and minute hands when the time is 2:15 a.m.?

7 Which is heavier: a pound of gold

or

a pound of feathers?

They aren't the same!

8 Which clock is better: one that loses a minute a day
                              or
                 one that does not run at all?

If the product $1 \times 2 \times 3 \times \cdots \times 28 \times 29 \times 30$ is factored into prime numbers,
how many 5s will the factorization contain?

> Knowing the no. of
> 5's tell us the no. of
> terminal zeros this
> product has.

10 Einstein regularly travels by plane between Singapore and Hong Kong. The journey lasts for 2.72 hours. If the plane takes off at 2:00 p.m., what time will he arrive at his destination?

Metamathematics in progress

$$3456710\ldots \times 745245\ldots$$
$$= 349782 \ldots 76\,3\,2\,3$$

Are numbers created or discovered?

Mathematical Quickies & Trickies

Casting Out Nines

The *casting out of nines* is an ancient and interesting method of checking the accuracy of an arithmetic computation. Any addition, subtraction, or multiplication can be checked by the method of casting out nines.

Consider the multiplication

$$33 \times 11 = 363.$$

To cast the nines out of 33, we add all the digits of the number, obtaining $3 + 3 = 6$. Similarly for 11 we get $1 + 1 = 2$. Casting nines out of 363, we see that $3 + 6 = 9$, so we cast out the 3 and the 6, leaving only the 3.

Now, instead of our original multiplication, we now have a simpler one:

$$6 \times 2 = 3, \text{ since } 6 \times 2 = 12, \text{ and } 1 + 2 = 3.$$

This shows that the original working was indeed correct—the "digital root" on both sides is 3.

Therefore $33 \times 11 = 363$.

Let's look at another example:

$$2959 \times 59 = 174{,}581$$

By casting out the two nines, the number 2959 becomes $2 + 5 = 7$. Similarly, 59 reduces to 5. Casting out nines or any combination of digits adding up to nine ($4 + 5$; $8 + 1$), the answer 174,581 reduces to 8.

Since the problem has reduced to $7 \times 5 = 35$, and $3 + 5 = 8$, the answer *must* be correct.

DANGER! Casting out 9's is not 100% foolproof!

When a problem has been checked by the method of casting out nines, and the two answers disagree, the problem is necessarily wrong. However, any agreement between the problem and the answer does not necessarily imply that the answer is correct or there is no error in the working.

Casting out nines is not a perfect method for detecting any error, no matter how gross. For example, is $32 \times 23 = 529$?

32×23 reduces to 5×5 or 25, and this gives $2 + 5$ which is 7. Casting out nines out of 529 gives $5 + 2$, or 7. There is an agreement between the problem and the answer, yet the method fails to detect the error, because $32 \times 23 = 736$.

More "casting out nines" on addition and subtraction

Example 1

To check whether $3459 + 2575 = 6034$.

Solution:

Cast out all 9's:
6 + 3
4 + 5, 9
2 + 7

Add the digits of the sum: $6034 \longrightarrow 4$

Add the digits of 3459: $(4 + 5)$; $9 \longrightarrow 3$ (casting out all the 9's)

Add the digits of 2575: $(2 + 7) \longrightarrow 5 + 5 = 10 \longrightarrow 1 + 0 = 1$

Check the final sums: $3 + 1 = 4$

Example 2

To check whether $24 \times 5.38 = 129.12$.

Thou shalt cast out thy nines!

Solution:

Add the digits of 129.12 (ignoring the 9s): $1 + 2 + 1 + 2 = \mathbf{6}$

Add the digits of 24: $2 + 4 = 6$

Add the digits of 5.38: $5 + 3 + 8 = 16 \longrightarrow 1 + 6 = 7$

Check the final digits: $6 \times 7 = 42 \longrightarrow 4 + 2 = \mathbf{6}$

1. Mentally calculate: 499,992 − 99,986.

Look for friendly numbers!

2. What are the next four numbers in this series?

12, 1, 1, 1, 2, 1, 3, ___, ___, ___, ___

Think laterally!

③ What is the product of the following series?

$$(x - a)(x - b)(x - c) \ldots (x - z)$$

Hard-looking questions often have simple answers!

④ At the half-way of a 1000-km race, Mr. Kiasu finds that he has been driving at an average speed of 50 km/h. How fast should he drive the second half of the race so as to achieve an overall average of 100 km/h?

⑤ To motivate her son to study mathematics, Mrs. Kiasu agrees to reward her son 50 cents for every problem solved correctly and to fine him 35 cents for each incorrect solution. At the end of 17 problems, neither owes anything to the other. How many problems did the boy solve correctly?

6 The Armed Forces promises the female officers $100 and a hamper as their rewards for a year. Ms. Singa leaves the service after 7 months, and receives the hamper and $20. How much is the hamper worth?

But do I need them?

7 Two cyclists leave town *A* for town *B* at the same time, where they stay 4 hours before returning to town *A*. John travels a speed of 30 km/h going and 40 km/h returning. Paul travels 35 km/h each way. Who gets back first?

Averagely speaking, both should arrive at the same time, but ...

8. A racing car covered a 6-km track at 140 km/h for 3 km, 168 km/h for 1.5 km, and 210 km/h for 1.5 km. What was the average speed covered for the 6 km?

I got F9 for my Math because I watched F1 the night before.

* 9. Flowing down with the current, a man took 4 min to cover 2 km, while against the current, it took him 8 min. How long would it take him to cover the same distance in still water (no current)?

Go with the current.

10   Is the number of people in the world who have shaken hands with an odd number of people odd or even?

Peace on Earth

10 Mathematical Quickies & Trickies

Be a Calculator Expert

Example 1

Each approximate decimal is obtained from a calculator by dividing a whole number by another. In each case, what are the two smallest possible whole numbers?

(a) 0.111111111 (b) 0.011111111 (c) 0.01010101

(d) 0.101010101 (e) 0.110011001

Solution:

(a) $0.1 = 1 \div 10$

But $0.111\ 1\ldots > 0.1$

Since $1/9 > 1/10$, so we try $1 \div 9$.

In fact, $0.111111111 = 1 \div 9$

(b) $0.011\ 111\ldots = 0.111\ 111\ldots \div 10 = \frac{1}{9} \div 10$

Clearly, $1 \div 90$ is the division.

Hence, $0.011111111 = 1 \div 90$

(c) $0.010\ 101\ldots < 0.011\ 111\ldots = \frac{1}{90}$

Since $\frac{1}{99} < \frac{1}{90}$, so we try $1 \div 99$.

Hence, $0.01010101 = 1 \div 99$

(d) $0.101\ 010\ldots < 0.111\ 111\ldots$

$0.101\ 010\ldots < 1 \div 9$ (or $10 \div 90$)

Since $\frac{10}{99} < \frac{10}{90}$, so we try $10 \div 99$.

Hence, $0.101010101 = 10 \div 99$

(e) 0.110 011 00 ... > 0.101 010 101 ...

0.110 011 00 ... > $\frac{10}{99} = \frac{100}{990}$

Since $\frac{10}{99} < \frac{100}{909}$, so we try $100 \div 909$.

Hence, $0.110011001 = 100 \div 909$

Example 2

Using the digits 1, 2, 3, 4, 5, 6 once only, make two numbers which multiplied together give the largest number possible.

Solution:

Using the calculator, and by trial and error, the largest possible product that can be formed from these six digits is

$$642 \times 531 = 340,902$$

Mathematical Mantra

MATH IS FUN
MATH IS COOL
MATH IS COOL
MATH IS FUN
MATH IS FUN
MATH IS COOL
MATH IS COOL
MATH IS FUN

1. On a calculator if you only use the "4" key and any of the +, −, ×, ÷ and = keys as many times as you like, what is the least number of presses you would need to make 9 on the calculator?

2. Three consecutive numbers multiplied together give 3360. What are the numbers?

3. Using the digits 1, 2, 3, 4, and 5 and one multiplication sign anywhere you like, what is the largest number you can make?

4 Sam divided a whole number by another whole number and the answer was 0.4705882 on his eight-digit calculator display. They were both less than 20. What are these two numbers?

5 Make 1000 with eight eights. You may use +, −, ×, ÷, and parentheses.

6 Using the digits of 1999 kept in that order and the operations +, −, ×, ÷, and parentheses, can you make any of the numbers from 1 to 20?
For example, 1 + (9 + 9) ÷ 9 = 3

7 To remember his boss's number, Ian begins "dialing" the number on his calculator to get into the habit. His number is 69432075. Last night after he dialed the number on his mobile phone, a voice at the other end of the line said, "Welcome, Saltville 63498015".

What happened and what did Ian do wrong?

---

8 *Antipedean Arithmetic*

Move just ONE line to change the subtraction sum such that it is then possible to read a correct sum.

$$182 - 882 = 695$$

The ten-digit number 4,103,987,526 is formed by 10 different digits. Enter this number in a calculator.

One way to delete the digit 0 is to proceed as follows:

4,103,987,526 − 4,100,000,000 + 410,000,000 = 413,987,526

To delete the digit 1, one may do the following:

413,987,526 − 410,000,000 + 40,000,000 = 43,987,526

Explain how would you delete the digits 2, 3, 4, and so on.

Use a calculator to find these multiplications, then predict what the next in the pattern will be when your calculator display is not big enough.

11 × 11
111 × 111
1111 × 1111
...

11 Using the digits 1, 2, 3, 4, 5, and 6 once each and two multiplication signs anywhere you like, what is the largest number possible?

12 Without using a calculator, which is smaller?

$$\frac{10^{11} + 1}{11^{10} + 1} \quad \text{or} \quad \frac{1 + 10^{12}}{1 + 10^{11}}$$

A Mind-Saving Abacus

In case of emergency, break open!

11 Mathematical Quickies & Trickies

Division by 9

Casting out nines may be used to check division by 9.

Divide the number by 9. Then, compare the remainder with the sum of the digits of the number (repeat until you obtain a one-digit number, called the *digital root*).

$$4754 \div 9 = 528\frac{2}{9} \text{ (the remainder is 2)}$$

$$4 + 7 = 11 \text{ (ignoring "5 + 4 = 9") and } 1 + 1 = 2$$

$$\text{Digital root} = 2; \text{Remainder} = 2$$

Note: The above method for division by 9 does not serve to check the accuracy of the quotient.

Warning: "Casting out nines" does not work for checking division by other numbers.

PRACTICE

1. Simplify the following divisions, and use casting out 9's to check the answers.

    (a) $\frac{96}{144}$ (b) $\frac{969}{5814}$ (c) $\frac{207}{23}$ (d) $\frac{783}{9}$

2. By casting out 9's, find the missing digit if 3_1,569 is divisible by 9.

3. By casting out 9's, find the remainder when 987,654 is divided by 9.

You may use a calculator to check your answers!

71

1 On a math whiz's tombstone,
   "Here lies Nerdo, born 3 B.C., died 4 A.D."
   How old was Nerdo when he died?

Do mortality and numeracy mix?

2 The alphabet has 26 letters. How many are not vowels?

Not 26 alphabets, but 1 alphabet with 26 letters.

3 John is 5 years old. Jane is 4 years old, and Jack is 8 years old. How many birthdays have all these children had?

4 Due to the many errors in a book, a printer had to replace pages 28, 29, 109, 130, and 131. How many sheets of paper did she have to reprint?

*⑤ How many triangles are there in this figure?

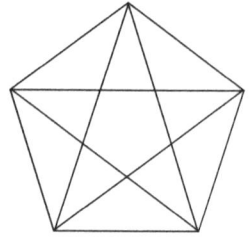

⑥ There is a child in front of two children, a child behind two children, and a child between two children. How many children were in the queue?

Queueing
Theory
4
Blockheads
&
Morons

7 A sheep rancher in New Zealand gets $1\frac{1}{2}$ pounds of wool from $1\frac{1}{2}$ sheep in $1\frac{1}{2}$ days. How many pounds can be obtained from 6 sheep in 7 days?

I prefer ratios to proportions!

8 There are 24 teenagers taking part in a knock-out tennis tournament with each winner moving to the next round. How many matches must be played before the cup winner emerges?

Solve this problem in more than 2 ways.

9. An ant starts crawling from the 12-cm end of a ruler along its edge. If it covers half the distance in 12 seconds, how much longer would it take the ant to reach the 1-cm mark?

10. A hotel custodian had to nail metal numbers on the doors. If the hotel rooms were numbered from 1 to 99, how many 7s are needed?

Ten Mathematical Trickies (Number Riddles)

1. A farmer had thirteen pigs. All but four died. How many were left?

2. **What number leaves nothing if you take away half?**

3. What occurs once in a minute, twice in a moment, but not at all in a second?

4. **Which is better: an old two-dollar bill or a new one?**

5. Seven is an odd number. How can you make it even?

6. **What is the difference between 100 and 1000?**

7. What time is it when the clock strikes thirteen?

8. **What can be measured, but has no length, width, or thickness?**

9. What has two hands and a face, but no arms or legs?

10. **A girl was 9 on her last birthday, and she will be 11 on her next. How is this possible?**

1. Four
2. 8
3. The letter m
4. An old $2 bill is better than a $1.
5. Take away the s.
6. Nothing (zero)
7. Time to get a new clock
8. Your temperature
9. A clock
10. Today is her 10th birthday.

1. If there are 12 one-cent stamps in a dozen, how many twenty-cent stamps are there in a dozen?

2. A machine has 25 blue balls and 25 red balls. Every one-dollar coin produces one ball from the machine. How many one-dollar coins must be spent before Joe could be certain of sharing two balls of the same color?

3 Mr. Yan went to a game shop and bought a Tangram set and a magazine. Both items cost $12.50. The Tangram set cost $2.50 more than the magazine. How much did each item cost?

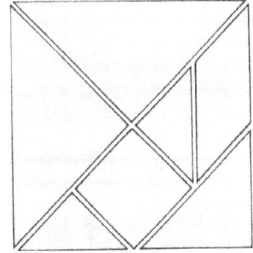

The 7-piece tangram

4 If 5 fishermen can catch 5 fish in 5 minutes, how long would it take for 50 fishermen to catch 50 fish?

5 How much dirt can be removed from a hole that is 5 cm deep, 3 cm wide, and 4 cm long?

What's the difference between 'capacity' and 'volume'?

6 Which is worth more: 4 pounds of $10 silver coins

or

2 pounds of $20 silver coins?

7 A 8-kg male donkey takes 80 minutes to cook, while a 4-kg female donkey takes 1 hour and 20 minutes to cook. What might account for the difference in cooking time?

8 A homemaker bought something that cost:

1	for	**$1.00**
12	for	**$2.00**
144	for	**$3.00**

Which item could reasonably account for this pricing?

9 When Susan was five, her mother measured her against a tree and marked it at a height of 3 feet. If the tree grows 1 foot every year, how high would the mark be after 12 years?

10 Five sections of chain are to be linked together to form a longer chain. Each section has three links. If it cost 5¢ to cut open a link and 10¢ to weld it shut, what would be the least amount of money it would cost to join the five sections?

$1 + 2 + 3 + \ldots + 24 + 25 = ?$

$1 + 2 = 3$
$3 + 3 = 6$
$6 + 4 = 10$
$\cdots$
$\cdots + 25 =$

$\dfrac{25 \times 26}{2} = 325$

Animal Calculators

13 Mathematical Quickies & Trickies

Ten Steps to be Math Smart

10. Take responsibility for your mathematical learning. Never depend solely on your mathematics teacher or tutor to gain mathematical knowledge and know-how.

9. Take part in extra-mathematical activities: solve nonroutine problems; subscribe to at least one mathematics magazine or newsletter to solve the problems of the week and month, attend math talks, watch math movies, and the like.

8. Have a healthy dose of mathematics every day: read expository articles about mathematics by borrowing books from school and public libraries.

7. Never faithfully accept any mathematical result. Ask questions when you do not understand something.

6. Accept the challenge of working with friends in acquiring mathematical concepts, and in improving your problem-solving skills.

5. Be willing to share your understanding of mathematics with others, and be open to the ideas of others.

4. Take pride in your work and never let yourself fall into the trap of believing you cannot do math.

3. Set high mathematical expectations for yourself; however, guard against mathematical pride and arrogance.

2. Be sensitive to the sight and sound of numbers—look for relationships that will provide insight and understanding.

1. Pray for divine wisdom to receive mathematical blessings — not only to gain mathematical knowledge by becoming numerate, but also to become mathematically civilized.

1. The difference in hours between the full-timers and the part-timers who work three hours a day is four hours. How long do full-timers work?

2. Mrs. Yan is given no more than $400 to spend on staying at a resort hut. What is the most number of days she can stay if the cost is $35 per day?

3. A liter of paint covers a surface area of 16 m². How much paint is needed to give two coats of paint to a ceiling 4.5 m by 5.5 m, if paint is sold in one-liter cans?

4 How many times can you subtract 2 from 17?

5 A factor of 60 is chosen at random.
   What fraction of all factors that
   (i) it is 20,
   (ii) it has factors of both 2 and 5?

   Any whole number
   is a factor of itself.

   True or False

6 How many 2-cent stamps are there in two dozens?

2¢

85

7. If it takes $12\frac{1}{2}$ minutes to boil an egg, how long does it take to boil 25 eggs?

8. John told Jane that if he had $1\frac{1}{2}$ oranges more than what he had he would have $1\frac{1}{2}$ times as many. How many oranges did John have?

9　A primary school plans to rent buses that can hold 37 students each. How many buses are needed if there are 640 students?

10　A man bought a camel for $500 and sold it later for $550. Then he bought it back for $600 and sold it for $650. How much did he make in the camel transaction?

Zero or Nothing?

That's zero.

0

And that's NOTHING.

Shortcuts à la Trachtenberg

The *Trachtenberg Method* was the brainchild of the Russian-born engineer, Jakow Trachtenberg, who devised shortcuts for everything from multiplication to algebra. He also came up with a method for teaching foreign languages, which is still used in many German schools today.

Trachtenberg was a political prisoner in Hitler's concentration camps—his name headed the dictator's most-wanted list. He organized the Society of Good Samaritans, and trained fellow Russians to care for the wounded.

To keep himself sane in jail, he kept busy by developing speedy methods to compute numbers. It is said that the calm logic of numbers were like old friends to him. To those who have not been exposed to unconventional methods of computing, the Trachtenberg system of shortcuts looks like some exhibition of mathematical wizardry — some "shorthand of mathematics."

Bribing the prison officers, Trachtenberg fled Germany to Switzerland, where he founded the Mathematical Institute in Zurich.

Let's illustrate how to multiply two numbers using the *Trachtenberg method*. Can you explain why it works?

Example 1

$$\times\, 11 \longrightarrow \text{Add each number to its neighbors.}$$

$$\frac{1234 \times 11}{1574}$$

$$\frac{2718 \times 11}{29898}$$

$$3 + 4 \qquad 2 + 3$$

$$1 + 8 \qquad 7 + 1 \qquad 2 + 7$$

Example 2

$\times 5 \longrightarrow$ Add half the neighbor; +5 if the number is odd.

$$\frac{4286 \times 5}{21430}$$

$\frac{1}{2} \times 6$

$\frac{1}{2} \times 8$

$\frac{1}{2} \times 2$

$\frac{1}{2} \times 4$

$$\frac{8628 \times 5}{43140}$$

$\frac{1}{2} \times 8$

$\frac{1}{2} \times 2$

$\frac{1}{2} \times 6$

$\frac{1}{2} \times 8$

$$\frac{5341}{26705} \times 5$$

$0 + \left(\frac{1}{2} \times 5\right)$
$= 2$

$\frac{1}{2} \times 1 = 0.5$

Take only the integer part.

$5 + \frac{1}{2} \times 3$
$= 5 + 1$
$= 6$
(5 is odd. Take the integer part of 1.5)

$5 + \left(\frac{1}{2} \times 4\right) = 7$
(3 is odd.)

Trachtenberg Method is like Mathemagic! Why does it work?

1. Write down the number "Eleven thousand eleven hundred eleven".

2. Bobby takes 3 minutes less than Betty to pack a box when each works alone. One day, after Bobby spent 6 minutes in packing a box, the teacher called him away, and Betty finished the job in 4 more minutes. How long would it take Bobby alone to pack a box?

3   Two boys were measuring a field on the banks of the Nile. Starting from one corner of the field, Ahmad walked 20 cubits south and Varu walked 60 cubits west. How far apart were they at this point?

4   One vegetable oil contains 6% saturated fats and a second 26% saturated fats. In making a dressing how many ounces of the second may be added to 10 ounces of the first if saturated fats is not to exceed 16%?

5   Water lilies double in area every 24 hours. At the start, there is only one water lily on the lake. If it takes 16 days for the lake to be completely covered with water lilies, on what day is the lake half covered?

6 What is the largest prime factor of 93,093?

7 The diagram shows five squares formed by arranging 16 matches.

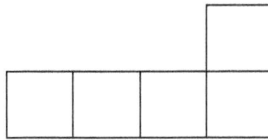

Move three sticks to form four squares.

8. A man is standing by the side of a river which is flowing past him at the rate of 6 km/h. He spots a raft 1 km upstream on which there is a boy shouting for help. He also spots two fishermen 1 km downstream paddling upstream to save him. If the two fishermen can paddle at a rate of 5 km/h in still water, how long will they take to reach the boy?

9. You just reach RingLand which is inhabited by two kinds of people: truth-tellers and liars. You meet two people and ask, "Are you truth-tellers or liars?" You can hardly hear the first person's reply. The second replies, "He says he is a truth-teller. So am I." Can you trust these two people for directing your journey?

10. When a 50-cent eraser found few buyers, its price was reduced. The remaining stock was sold for $31.93. What was the reduced price?

15 Mathematical Quickies & Trickies

Geometrical Quickies 1

Example 1

A triangle has sides 6, 8, and 10.
Find the radius of the inscribed circle.

Solution:

Label the sides a, b, and r.

$2a + 2b + 2r = 6 + 8 + 10 = 24$

Since $a + b = 10$, $2a + 2b = 20$

Thus, $2r = 4$

Hence the radius, r, of the circle is 2 units.

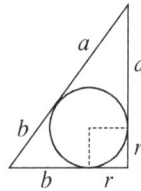

Example 2

How many slices are needed to saw the cubical block into
64 smaller cubes?

Solution:

The first cut slices the block in half. Then put the two pieces
on top of each other, and cut them into 4 slabs.

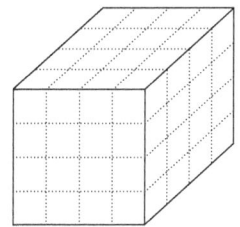

The third cut gives 8 slices. Then line up all 8 half-slices and cut them into
16 four-by-one slices.

The last two cuts will halve the pieces twice into 64 cubes.
Hence, six slices are needed to saw the block into 64 smaller cubes.

1. A 6-storey apartment has stairs of the same length from floor to floor. How many times is a climb from the first to the sixth floor higher than a climb from the first to the third floor?

2. Insert a mathematical symbol between 7 and 8 to express a number greater than 7 and less than 8.

3. What three whole numbers whose sum equals their product?

④ *ABCD* are four consecutive digits in increasing order.
DCBA are the same four in decreasing order.
The four *X*s represent the same four digits in an unknown order.
If the sum is 12,300, what number is represented by the four *X*s?
(W. T. Williams & G. H. Savage)

$$
\begin{array}{r}
A\,B\,C\,D \\
D\,C\,B\,A \\
X\,X\,X\,X \\
\hline
1\,2\,3\,0\,0
\end{array}
$$

⑤ How many different 10-digit numbers, such as 4,207,168,539, can be written by using all 10 digits?

⑥ At Jack's Cafe, the coffee served is 97% caffeine-free. How many cups of coffee would one have to drink to get the amount of caffeine in a cup of regular coffee?

7. How many squares of all sizes are there on the checkerboard?

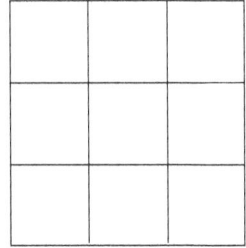

8. March has five Fridays. Three of them fall on even-numbered dates. What is the date of the fourth Friday in March?

The number 13 is most likely to fall on a Friday than on any other day of the week.

9   If PLUS equals 68, what does MINUS equal?

   A. 102   **B.** 76   **C.** 120   **D.** 46

10   Which is better: to get 5% of seven billion or 7% of five billion?

Did you know...

For Americans, one billion is one thousand million.

For British, one billion is one million million.

16 Mathematical Quickies & Trickies

Geometrical Quickies 2

Example 1

Given that XY = 14 cm, find the area of the shaded region, expressing your answer in terms of π.

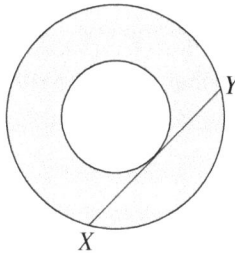

Solution:

Imagine the two circles decreasing in sizes, with the chord XY remaining 14 cm long, until the inner circle becomes a point circle.

Its radius would then become zero, and the circle would simply be the point which is the center of the larger circle.

The chord XY has now become the diameter of the larger circle and its area is thus $\pi \times 7^2 = 49\pi$ cm^2.

Compare the above method (which uses the *Zero Option*) with one that uses an algebraic method.

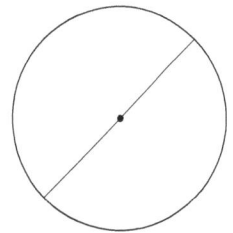

> For more on the Zero Option, zero in on K C Yan's *Geometrical Quickies & Trickies*.

1. Why would a barber in Malaysia rather cut the hair of two Singaporeans than that of one Japanese?

2. Each of the two equal sides of an isosceles triangle is 1 cm long. Find the length of the third side that will give the greatest area.

3. Find two whole numbers, x and y, such that

$$\text{H.C.F. } (x, y) \times \text{L.C.M. } (x, y) = xy.$$

Did you know...

What the British call the H.C.F., the Americans call the G.C.D. (Greatest Common Divisor).

④ In MathLand a date such as April 5, 1998, is often written 4/5/98, but in other places the month is given second and the same date is written 5/4/98. If you are not told which system is being used, how many dates are ambiguous in this two-slash notation? (David L. Silverman)

⑤ Regular hexagons are inscribed in and circumscribed outside a circle. If the smaller hexagon has an area of 3 square units, what is the area of the larger hexagon? (Charles W. Trigg)

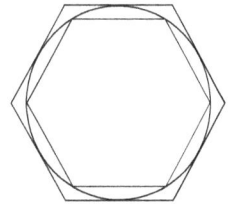

(6) "I will be x years old in the year x^2," said Ian in 1987. When was he born?

(7) How many 2-digit whole numbers are there from 1 to 222?

(8) To motivate her son to study mathematics, a mother promises to pay him 50¢ for every problem correctly solved, and to fine him 30¢ for each incorrect solution. At the end of 40 problems, neither owes anything to the other. How many problems did the boy solve incorrectly?

> **The Mathematics of Kiasuism**
>
> Bad deal:
> **Money + Time + Effort > Value**
>
> Good deal:
> **Money − Effort = Full Value**
>
> Kiasuism: 1 + 1 > 2
>   or
>   2 − 1 > 1

9 By what fractional part does four-fourths exceeds three-fourths?

10 Pampers has always celebrated her birthday with a cake decorated with the number of candles matching her age. She has, so far, blown off 190 candles. How old is Pampers?

17 Mathematical Quickies & Trickies

Geometrical Quickies 3

Example 1

A circle of radius 10 cm intersects another circle of radius 15 cm at right angles. What is the difference of the areas of the unshaded regions, leaving your answer in terms of π?

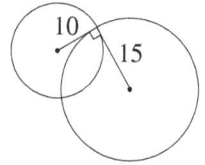

Solution:

If the two areas X and Y have a common area Z, then the unshaded areas are $(X - Z)$ and $(Y - Z)$.

The difference of the areas of the unshaded regions is $[(X - Z) - (Y - Z)] = X - Y$.

The difference is $\pi(15)^2 - \pi(10)^2 = 125\pi$ cm^2.

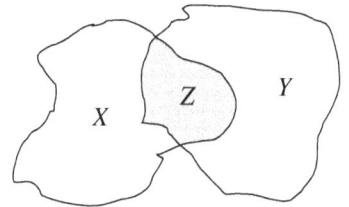

Example 2

The figure shows a square containing two quarter circles inside it. Find the difference between the two shaded areas. (Take π to be $\frac{22}{7}$.)

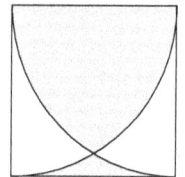

7 cm

Solution:

7 cm   7 cm

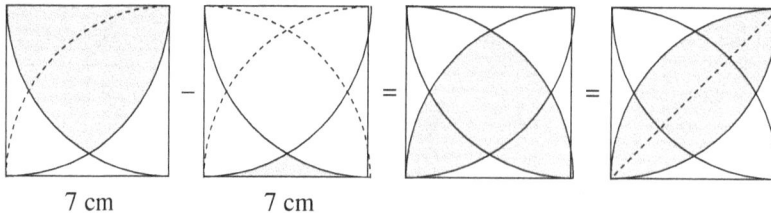

Difference between the two shaded areas
$$= 2 \times \left[\left(\frac{1}{4} \times \frac{22}{7} \times 7 \times 7\right) - \frac{1}{4} \times 7 \times 7\right] = 28 \text{ cm}^2$$

① Mentally calculate: 99,996 + 99,988.

② Find the remainder when the number

$$1\ 2\ 3\ 4\ 5\ 6\ 7\ 8\ 9\ 8\ 7\ 6\ 5\ 4\ 3\ 2\ 1$$

is divided by 9.

What's the divisibility test for 9?

3 Given that a and b are two distinct positive numbers, which is greater,

$$a^2 + b^2 \text{ or } 2ab?$$

4 What is the chance that, if you throw two dice, you will get at least one 4, 5, or 6?

5 What is the smallest number divisible by 11 which, when divided by any of the numbers from 2 to 10 inclusive, leaves a remainder of 1?

6 Mr. Watts sold his last two second-hand books at the price of $18 each. He believed he must have made a profit on the two sales since he made a 50% profit on one and only took a 28% loss on the other. Was Mr. Watts correct in his estimate?

Thou shalt not trust percentages!

7 What is the smallest possible whole number that gives the same non-zero remainder when divided by 3, 5, 7, or 11?

8 Find the number of zeros at the end of the product

$$1 \times 2 \times 3 \times \cdots \times 98 \times 99 \times 100.$$

9 A given prime number is the sum of two other prime numbers. Given that the sum of the two largest primes is 1200, find the three prime numbers.

Prime numbers are the 'atoms of mathematics'!

10 At a Grand Prix, one spectator observed that one quarter of the cars in front of last year's champion added to five-sixths of those behind him gives the number of cars running in the race. How many cars were running in the Grand Prix?

Singapore Grand Prix
(under the stars)

18 Mathematical Quickies & Trickies

(sugar + coffee) + milk = sugar + (coffee + milk)

For any three numbers *a*, *b*, and *c*: $(a + b) + c = a + (b + c)$

For example, $(2 + 3) + 4 = 2 + (3 + 4) = 9$

The above law is called the *Associative Law of Addition*.

Some non-associative phrases are:

1. He is in the [primary (school building)].
    He is in the [(primary-school) building].

2. James is unaware how [good (study is)].
    James is unaware how [(good study) is].

3. Lucifer stays into the [dark (brown room)].
    Lucifer stays into the [(dark brown) room].

> Do you know the difference in each case?

Applications of Associative Law

Example 1

Add $598 + 2367 + 402$ without paper and pencil.

Solution:

$$598 + 2367 + 402 = (598 + 402) + 2367$$
$$= 1000 + 2367$$
$$= 3367$$

Note that it is not easy to add 598 and 2367. But if we add 598 to 402, we get 1000. It is now easier to add 2367 to 1000.

Example 2

Add 345 + 15,999 without paper and pencil.

Solution:

$$345 + 15{,}999 = (344 + 1) + 15{,}999$$
$$= 344 + (1 + 15{,}999)$$
$$= 344 + 16{,}000$$
$$= 16{,}344$$

Like addition, multiplication is also associative.

For any three numbers a, b, and c: $\boxed{(a \times b) \times c = a \times (b \times c)}$

For example, $(2 \times 3) \times 4 = 2 \times (3 \times 4) = 24$

In short, $(ab)c = a(bc)$.

The above law is called the *Associative Law of Multiplication.*

Example 3

Perform the following: (a) $25 \times 78 \times 4$, (b) $125 \times 37 \times 8$.

Solution:

(a) $25 \times 78 \times 4 = (25 \times 4) \times 78$
$$= 100 \times 78$$
$$= 7800$$

(b) $125 \times 37 \times 8 = (125 \times 8) \times 37$
$$= 1000 \times 37$$
$$= 37{,}000$$

It is easier to look for partial product of numbers, whose value equals some powers of 10. Then multiply the product by the remaining numbers.

1. Mentally calculate: (a) 136×125,
   (b) 256×375.

2. A watch is set correctly at 4:00 p.m. It loses 2 minutes every hour. What is the correct time when the watch reads 6:30 a.m. the next day?

3. What is the next number in this sequence?

$$2, 12, 30, 56, 90, \underline{\quad}$$

④ A fireman who was standing on the middle rung of a ladder moved up 4 rungs. Due to the heat, he moved down 7 rungs. Two minutes later, he climbed another 5 rungs and remained there until the fire was put off. He then went up 6 more steps before he entered the building. How many rungs did the ladder contain?

⑤ What is the chance that, if you throw a pair of dice three times in succession, you will get at least one 4, 5, or 6 each time?

6 A priest left his possessions to be shared among his three sons in the proportions: $\frac{1}{2}$, $\frac{1}{3}$, $\frac{1}{9}$. After the taxes were deducted, $17,000 was left. How much did each son receive?

> Thou shalt divide my will proportionately. The angels are my witnesses!

*7 Mr. Yan is a bookworm. His wife grumbles that if he keeps on buying more books at the present rate, he will soon have over 50 of them. Twenty percents of his books are recreational and one-seventh were given by his friends. How many books does Mr. Yan have on his shelf?

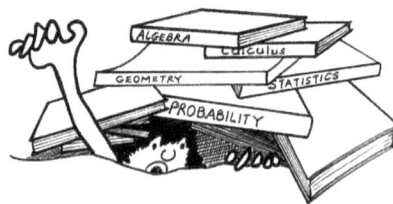

8　Three hundred mathletes participated in the last mathematics contest. Twelve percents of them brought one calculator. Of the remaining, half brought two calculators, and half brought none. How many calculators were there?

I cost more than a four-function calculator

*9　A swimmer can swim at twice the speed of the prevailing current. It takes him a total time of four minutes to swim from one end of the river bank and back again. How long would it take him to swim in still water?

10 How many deer could jump over a fence in an hour if 10 deer take 10 minutes to jump over a fence?

Santa's beloved deer

19 Mathematical Quickies & Trickies

More Applications of Number Laws

Example 1

Mentally compute the following using the appropriate laws.

(a) 17 + 11 + 23 + 49

Solution:

$$17 + 11 + 23 + 49 = (17 + 23) + (11 + 49)$$
$$= 40 + 60$$
$$= 100$$

(b) 51 + 114 + 69 + 186

Solution:

$$51 + 114 + 69 + 186 = (51 + 69) + (114 + 186)$$
$$= 120 + 300$$
$$= 420$$

(c) 8 + 45 + 28 + 64 + 55

Solution:

$$8 + 45 + 28 + 64 + 55 = (8 + 28 + 64) + (45 + 55)$$
$$= (36 + 64) + 100$$
$$= 100 + 100$$
$$= 200$$

(d) 7 + 998 + 63

Solution:

$$7 + 998 + 63 = (7 + 63) + (1000 - 2)$$
$$= (70 + 1000) - 2$$
$$= 1070 - 2$$
$$= 1068$$

Example 2

Apply the number laws to evaluate the following.

(a) $5 \times 31 \times 3 \times 2$

Solution:

$$5 \times 31 \times 3 \times 2 = (5 \times 2) \times (31 \times 3) \qquad \textit{Associative Law}$$
$$= 10 \times 93$$
$$= 930$$

(b) $35 \times 80 \times 2$

Solution:

$$35 \times 80 \times 2 = (35 \times 2) \times 80 \qquad \textit{Associative Law}$$
$$= 70 \times 80$$
$$= 5600$$

(c) 56×25

Solution:

$$56 \times 25 = \left(56 \times \frac{1}{4}\right) \times (4 \times 25) \qquad \textit{Law of One}$$
$$= 14 \times 100$$
$$= 1400$$

(d) 125×136

Solution:

$$125 \times 136 = \left(8 \times 125\right) \times \left(136 \times \frac{1}{8}\right) \qquad \textit{Law of One}$$
$$= 1000 \times 17$$
$$= 17,000$$

① Compute the following expressions.
(a) $1 + 2 + 3 + \cdots + 999 + 1000$
(b) $9999 + 9998 + \cdots + 3 + 2 + 1$
(c) $1 - 2 + 3 - 4 + 5 - 6 + 7 - 8 + \cdots + 4999 - 5000$

② From the numbers 1 to 100 inclusive, which of the digits 0 to 9 appears least, and which appears most?

3 There are 361 undergraduates taking statistics this year and each class has the same number of students. How many classes are there?

4 If there are 100 seconds in a minute, how many minutes are there in one hour?

* 5 What is the chance that, if you throw three dice, at least one of them will be a six?

6　If the cold tap alone is turned on full, a bathtub can be filled in 3 min. If the hot tap alone is turned on full, the bathtub can be filled in 4 min. When the plug is pulled out, the bathtub can be emptied in 2 min. How long will it take for the bathtub to be filled with both taps turned on full, but the plug pulled out?

7　There are $9 \times 8 \times 7 \times ... \times 3 \times 2 \times 1 = 362,880$ different ways of forming a nine-digit number using the digits 1 to 9 exactly once. What percentage of these are prime numbers, correct to the nearest 1%?

8. Find the greatest whole number that divides 212, 260, and 308, and leaves the same remainder in each case.

9. Each packet of *Mathties* contains the same number of candies. Last month I bought several packets of *Mathties* consuming 301 candies. This month my craving led me to consume another 516 additional candies. How many packets did I buy altogether, and how many candies are there in each packet?

*10. In a day, how many times does the hour hand point in the same direction as the minute hand?

122

Law of One: *The product of a number multiplied by 1 is that number.*

Applications of the Law of One

Example 1

$$25 \times 36 = (4 \times 25 \times 36) \times \frac{1}{4}$$
$$= (4 \times 25) \times \left(36 \times \frac{1}{4}\right)$$
$$= 100 \times 9$$
$$= 900$$

Example 2

$$2\frac{1}{8} \times 48 = \left(8 \times 2\frac{1}{8}\right) \times \left(48 \times \frac{1}{8}\right)$$
$$= \left(8 \times \frac{17}{8}\right) \times 6$$
$$= 17 \times 6$$
$$= 102$$

Example 3

$$33\frac{1}{3} \times 15 = \left(3 \times 33\frac{1}{3}\right) \times \left(15 \times \frac{1}{3}\right)$$
$$= \left(3 \times \frac{100}{3}\right) \times 5$$
$$= 100 \times 5$$
$$= 500$$

① A man was born in the year 1180 and yet died in the year 1163. Using the normal calendar, how is this possible?

② How many cubes of various sizes are there in a cube made of $5 \times 5 \times 5$ unit cubes?

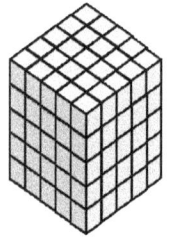

③ Which solid has the most volume?

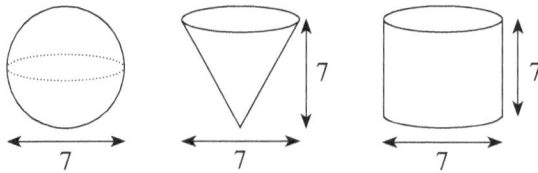

④ Ravin left Town *A* for Town *B*, traveling at a constant speed. When one third of the journey was over, the time was 11:12 a.m. When one quarter of the journey remained to be covered, the time was 12:32 p.m. When did Ravin reach Town *B*?

⑤ To number the pages of a Maths book, 660 digits (0, 1, 2, ..., 9) were needed. How many pages does the book contain?

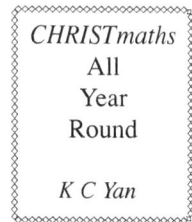

CHRISTmaths
All
Year
Round

K C Yan

125

6　Red bulbs flash every 8 seconds, yellow bulbs every 10 seconds, and green bulbs every 15 seconds. If they start flashing at the same time, after how long will they flash together again the fifth time?

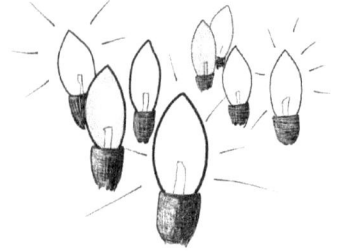

7　What is the smallest possible whole number which, when divided by 2, 3, 4, 5, 6, and 7 will give remainders of 1, 2, 3, 4, 5, and 6 respectively?

8　The figure shows 13 matches representing 4 squares.

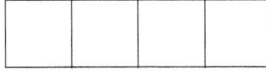

By removing one of the matches, arrange the remaining 12 matches such that they create five squares in total.

9　Farmer Bill keeps goats, cows, and chickens in the same enclosure. There are twice as many chickens as there are cows. He found that there are altogether 18 heads and 52 legs among his family of animals. How many goats are there in his enclosure?

*⑩ Mary bought a certain number of chocolate bars for $11.05. She sold each bar at 80 cents for a profit. Given that each chocolate bar costs more than 50 cents, how much money did she make?

Triskaidekaphobia

21 Mathematical Quickies & Trickies

Distributive Law

Given any three numbers a, b, and c,

$$a \times (b + c) = (a \times b) + (a \times c)$$

or

$$(b + c) \times a = (b \times a) + (c \times a)$$

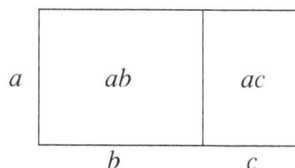

e.g., $2 \times (3 + 4) = (2 \times 3) + (2 \times 4) = 6 + 8 = 14$

Check: $2 \times (3 + 4) = 2 \times 7 = 14$

Another example: $4 \times (5 - 2) = (4 \times 5) - (4 \times 2) = 20 - 8 = 12$

Check: $4 \times (5 - 2) = 4 \times 3 = 12$

Caution:

$$5(x + 1) \neq 5x + 1 \text{ but } 5(x + 1) = 5x + 5$$

Note:

$$(a - b) \cdot \boxed{m - n} = a \cdot \boxed{m - n} - b \cdot \boxed{m - n}$$
$$= am - an - bm + bn$$

More examples

$$[(-3) + (-4)] \times (+5) = [(-3) \times (+5)] + [(-4) \times (+5)]$$
$$= (-15) + (-20)$$
$$= -35$$

$$(-3) \times [4 + (-5)] = [(-3) \times 4] + [(-3) \times (-5)]$$
$$= (-12) + (15)$$
$$= 3$$

Applications of Distributive Law

Examples

1. $19 \times 39 = (20 - 1) \times 39$
   $= (20 \times 39) - (1 \times 39)$
   $= 780 - 39$
   $= 741$

   $19 \times 39 = 19 \times (40 - 1)$
   $= 19 \times 40 - 19$
   $= 760 - 19$
   $= 741$

2. $15 \times 68 = (10 + 5) \times 68$
   $= (10 \times 68) + (5 \times 68)$
   $= 680 + \left(\frac{10}{2} \times 68\right)$
   $= 680 + 340$
   $= 1020$

   $15 \times 68 = 15 \times 2 \times 34$
   $= 30 \times 34$
   $= 1020$

3. $5\frac{3}{7} \times 28 = \left(5 + \frac{3}{7}\right) \times 28$
   $= (5 \times 28) + \left(\frac{3}{7} \times 28\right)$
   $= 140 + 12$
   $= 152$

   > How do I use the distributive law to show that $(-2) \times (-3) = (+6)$?

4. $\frac{1}{6} \times 19 + \frac{1}{6} \times 35 = \frac{1}{6} \times (19 + 35)$
   $= \frac{1}{6} \times 54$
   $= 9$

5. $179 \div 9 = \frac{1}{9} \times (162 + 17)$
   $= \left(\frac{1}{9} \times 162\right) + \left(\frac{1}{9} \times 17\right)$
   $= 18 + 1\frac{8}{9}$
   $= 19\frac{8}{9}$

   $179 \div 9 = (180 - 1) \div 9$
   $= \frac{1}{9} \times 180 - \frac{1}{9} \times 1$
   $= 20 - \frac{1}{9}$
   $= 19\frac{8}{9}$

1. At Yan's MathClub, satay costs 35 cents per stick. This is 2 cents per stick less than a stick at Goh's MathClub. If you need 4 sticks of satay, how much will you pay at Goh's?

2. Which fraction is larger: $\frac{2009}{2010}$ or $\frac{2010}{2011}$?

No LCM, please!

3 How many 3 cm × 3 cm squares can be cut from a cardboard measuring 36 cm × 16 cm?

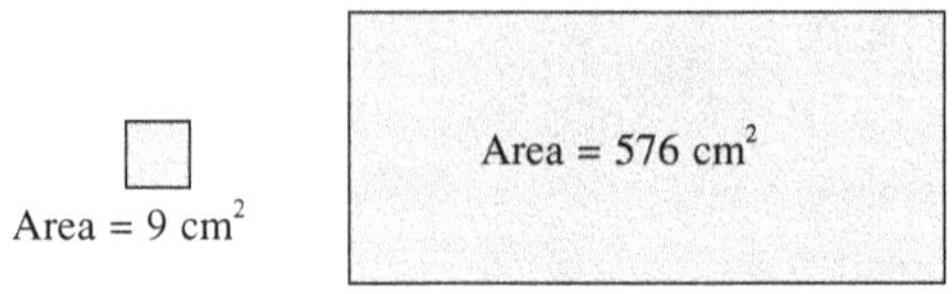

Area = 9 cm^2

Area = 576 cm^2

4 Evaluate $\dfrac{1}{1 \times 2} + \dfrac{1}{2 \times 3} + \dfrac{1}{3 \times 4} + \dfrac{1}{5 \times 6} + \ldots + \dfrac{1}{2009 \times 2010}$.

⑤ Tim bought 20 stamps featuring Pascal at $\$x$ each and 30 stamps featuring Pythagoras at $\$y$ each. He bought the entire collection for $\$340$. If a Pythagoras-stamp costs 5 times more than a Pascal-stamp, calculate how much did he pay for the Pascal-stamps?

⑥ (a) How many whole numbers containing the digit 2 are there between 100 and 500?

(b) How many times does the digit "2" appear between 100 and 500?

* **7** Find the smallest positive whole number that has a remainder of 1 when divided by 7, and a remainder of 3 when divided by 11.

8 A car traveled a total distance of 20,000 km. If five tires were rotated regularly to save wear, how many kilometers did each tire travel for the entire trip?

9 Given that x, y, z are whole numbers such that each is the product of the other two, find the values of x, y, and z.

* 10 In a refugee camp 20 pounds of rice are to be distributed among 20 men, women, and children. Each man, woman, and child is allocated 3 pounds, $1\frac{1}{2}$ pounds and $\frac{1}{2}$ pound, respectively. How many men, women, and children are there?

22 Mathematical Quickies & Trickies

More Applications of Distributive Law

Example 1

Compute 1001×102 without pencil and paper.

Solution:

$$
\begin{aligned}
1001 \times 102 &= 1001 \times (100 + 2) \\
&= 1001 \times 100 + 1001 \times 2 \\
&= (1000 + 1) \times 100 + (1000 + 1) \times 2 \\
&= 100{,}000 + 100 + 2000 + 2 \\
&= 102{,}102
\end{aligned}
$$

Example 2

Show how 9×58 may be computed in more than one way.

Solution:

Method 1

$$
\begin{aligned}
9 \times 58 &= 9 \times (50 + 8) \\
&= (9 \times 50) + (9 \times 8) \\
&= 450 + 72 \\
&= 522
\end{aligned}
$$

Method 2

$$
\begin{aligned}
9 \times 58 &= 9 \times 29 \times 2 \\
&= 261 \times 2 \\
&= 522
\end{aligned}
$$

Method 3

$$
\begin{aligned}
9 \times 58 &= 9 \times (60 - 2) \\
&= (9 \times 60) - (9 \times 2) \\
&= 540 - 18 \\
&= 522
\end{aligned}
$$

Method 4

$$
\begin{aligned}
9 \times 58 &= (10 - 1) \times 58 \\
&= (10 \times 58) - (1 \times 58) \\
&= 580 - 58 \\
&= 522
\end{aligned}
$$

Example 3

Prove that $(-1) \times (-1) = +1$.

Solution:

$$0 = -1 \times 0$$
$$0 = -1 \times [(+1) + (-1)]$$
$$0 = (-1) \times (+1) + (-1) \times (-1) \qquad \textit{Distributive law}$$
$$0 = -1 + (-1) \times (-1)$$
$$\text{Thus } (-1) \times (-1) = +1$$

Example 4

Without pencil and paper, compute the following.

(a) 1001×55 (b) 997×45 (c) $4\frac{2}{7} \times 21$

(d) 15×66 (e) $153 \div 7$

Solution:

(a) $\begin{aligned}[t] 1001 \times 55 &= (1000 + 1) \times 55 \\ &= 1000 \times 55 + 1 \times 55 \\ &= 55{,}000 + 55 \\ &= 55{,}055 \end{aligned}$

(b) $\begin{aligned}[t] 997 \times 45 &= (1000 - 3) \times 45 \\ &= 1000 \times 45 - 3 \times 45 \\ &= 45{,}000 - 135 \\ &= 44{,}865 \end{aligned}$

(c) $\begin{aligned}[t] 4\frac{2}{7} \times 21 &= \left(4 + \frac{2}{7}\right) \times 21 \\ &= (4 \times 21) + \left(\frac{2}{7} \times 21\right) \\ &= 84 + 6 \\ &= 90 \end{aligned}$

(d) $\begin{aligned}[t] 15 \times 66 &= (10 + 5) \times 66 \\ &= (10 \times 66) + (5 \times 66) \\ &= 660 + \frac{660}{2} \\ &= 990 \end{aligned}$

$\begin{aligned}[t] 15 \times 66 &= 30 \times 33 \\ &= 990 \end{aligned}$

(e) $\begin{aligned}[t] 153 \div 7 &= \frac{1}{7} \times (140 + 13) \\ &= \left(\frac{1}{7} \times 140\right) + \left(\frac{1}{7} \times 13\right) \\ &= 20 + 1\frac{6}{7} \\ &= 21\frac{6}{7} \end{aligned}$

1 Find the remainder of $7 \div 0.4$

2 Evaluate $-(-x - x^x)^{-x}$ if $x = -2$.

3 Evaluate $\left(1 - \frac{1^2}{100}\right) \times \left(1 - \frac{2^2}{100}\right) \times \left(1 - \frac{3^2}{100}\right) \times ... \times \left(1 - \frac{2012^2}{100}\right)$.

4   In a certain univerrsity there are 6 times as many students as professors. Write an equation that describes the relation between the number of students (S) and the number of professors (P).

* 5   To prevent terrorists from sabotaging the phone lines of members of Parliament, the CID is planning to connect every member directly to every other member with one long phone line. Ignoring wiring complications, how many lines would be needed to implement this project if 36 members are involved?

WANTED

Ozahma bin Salamat

**$1,000,000
Reward**

6　Ten masons can dig 20 holes in 40 days. Working at the same rate, how many days can 20 masons dig 10 holes?

*7　A chessboard has 8 by 8 squares. How many individual rectangles are there altogether?

8　A mathematics counsellor post drew 100 applicants. 20 did not pass their English and Mathematics. 55 were good in English, and 75 were good in Mathematics. How many applicants were qualified for the post, which requires someone to be proficient in both English and Mathematics?

Literacy
+ Numeracy
—————
Good Job

* **9** What is the least whole number that leaves remainders 1, 3, and 1 when divided by 3, 5, and 7 respectively?

10 Samuel has won three games consecutively. What are his chances, in percentages, of winning the fourth game?

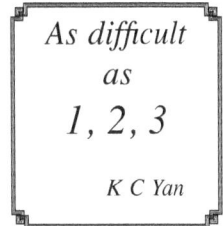

As difficult as

1, 2, 3

K C Yan

2018

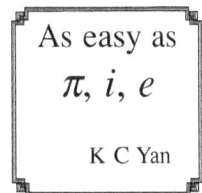

As easy as

π, i, e

K C Yan

2020

Mathematical Quickies & Trickies

The Joy of Guesstimation

Some remedies to combat number numbness involve indulging in some estimation prestidigitation. For example, we could "guesstimate" ("guess" + "estimate") to detect the reasonableness of any calculated figures or answers.

Addition Guesstimation

$$
\begin{array}{r}
6356 \\
+ \quad 7823 \\
\hline
14\ 179
\end{array}
\quad \approx \quad
\begin{array}{r}
6000 \\
+ \quad 8000 \\
\hline
14\ 000
\end{array}
\quad \text{(Round off to the nearest thousand)}
$$

Something big

$$
\begin{array}{r}
43\ 765\ 315 \\
+ \quad 8\ 427\ 562 \\
\hline
52\ 192\ 877
\end{array}
\quad \approx \quad
\begin{array}{r}
44\ 000\ 000 \\
+ \quad 8\ 000\ 000 \\
\hline
52\ 000\ 000
\end{array}
\quad \text{or} \quad
\begin{array}{r}
43.8 \text{ million} \\
8.4 \text{ million} \\
\hline
52.2 \text{ million}
\end{array}
$$

Multiplication Guesstimation

$$
\begin{array}{r}
76 \\
\times \quad 43 \\
\hline
3268
\end{array}
\quad \approx \quad
\begin{array}{r}
80 \\
\times \quad 40 \\
\hline
3200
\end{array}
$$

Subtraction Guesstimation

$$
\begin{array}{r}
5358 \\
- \quad 3713 \\
\hline
1645
\end{array}
\quad \approx \quad
\begin{array}{r}
5000 \\
- \quad 4000 \\
\hline
1000
\end{array}
\quad \text{or} \quad
\begin{array}{r}
5400 \\
- \quad 3700 \\
\hline
1700
\end{array}
$$

Division Guesstimation

$$
\begin{array}{r}
5\ 473.6 \\
8\,\overline{)43\ 789}
\end{array}
\quad \approx \quad
\begin{array}{r}
5 \\
8\,\overline{)44\ 000} \\
40 \\
\hline
4
\end{array}
\qquad \frac{4}{8} = \frac{1}{2} = 0.5, \text{ so } \frac{43\ 789}{8} \approx 5.5 \text{ thousands}
$$

1. Mr. Barn saw 12 sparrows. He shot three of them. How many birds remained?

2. Grandma cut the cake into 12 equal pieces and took five seconds to make each cut. If she started cutting at 3:45 p.m., when would she finish?

* 3. A *palindromic time* is one that reads the same from backwards and forwards, such as 02:55:20, as displayed on a 24-hour digital. In a day, how many times will the digital clock display a palindromic time?

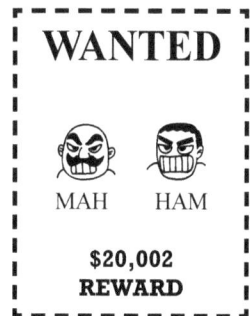

WANTED

MAH   HAM

$20,002
REWARD

12:21

④ At the beginning of a trip, a car's meter registered 9909. At the end of the trip, the meter registered 0109. How many kilometers were traveled on this trip?

⑤ A shoplifter in a coin dealer's shop stole the oldest coin dated 156 B.C. If a rare coin is worth $20 for each year before Christ that it was minted, how much could he sell it for?

144

6. Mr. Kiasu was stranded on a desert island with just one box of 64 cigars. He wanted his smoking habit to last as long as possible. He found that by smoking exactly $\frac{3}{4}$ of each cigar, he could stick four butts together to make another whole cigar. How many extra cigars can he smoke?

* 7. There are only two rectangles whose areas are numerically equal to their perimeters. Find the whole numbers dimensions of the rectangles.

8 At what time between 7 o'clock and 8 o'clock will the two hands of the clock be in a straight line?

9 If 6 boys can move 6 boxes in 6 min, how many boys are required to move 60 boxes in 60 min?

10 Find the largest whole number which can be written with four 1s.

1   **1**   **1**   **1**

24 Mathematical Quickies & Trickies

Are You a *Fermi Disciple*?

Let's look how you could solve some seemingly impossible quantitative problems, what mathematicians and scientists called the *Fermi problems*.

> ***How many trees are felled each year to print the textbooks?***
> *How many teenagers smoked cigarettes every year? (How many zeros?)*
> ***How many grains of sand are there at the Sentosa beach?***
> How many water taps are there in Singapore?

The above problems might appear impossible to compute, but not for the physicist Enrico Fermi (1901–1954), who liked to amuse his listeners by posing such problems.

The gist of a Fermi problem is that a well-informed person can solve it using a series of estimates yielding a reasonably good approximate answer. All you need is some known or estimable data and common sense.

Dr. Fermi won the Nobel Prize in 1938 when he was 37.

Physicists and mathematicians solve Fermi problems every day. Given a Fermi problem, one can come up with different methods for generating independent estimates. The results can then be compared to ensure reasonable conclusions.

You might feel challenged when faced with unfamiliar problems, especially if they look strange and they don't have an algorithm. For many non-mathematicians, mathematics connotes exactitude. Solving Fermi problems is not only a good way to get you loosen up by getting some good, if dirty, results, but also helps you to enhance your estimation skills.

You might not care how many stars there are in the universe, how high a billion of coins stacked together is, or whether the number of lies each day outnumbers the number of wrong motives. But solving Fermi problems provides an indirectly good opportunity to use many techniques that are useful in solving all kinds of quantitative problems.

Moreover, the nicest Fermi problems are intellectually stimulating although the challenges might be artificial or impractical.

The dexterity to solve Fermi problems testifies that problem-solving ability is often limited not by incomplete information but by the inability to use the information that is already available.

You needn't be a Fermi-problem specialist. Too often, all you need is an order-of-magnitude estimate to detect whether the numbers are reasonable or ridiculous. The more conscious and enthusiastic you become of Fermi problems, the more confident you will be at estimating events.

Here are some guesstimation exercises to stimulate or tickle your brain.

1. **How much garbage is accumulated by Singapore residents every year?**

2. Estimate the thickness of a book that contains a million letters.

3. **How many millions of dollars collected from tuition fees are undeclared every year?**

4. A sheet of paper 0.01 cm is folded fifty times in succession. How thick is the resulting wad of paper?

5. What is the total length of all the hair on an average woman's head?

1 Which 3-digit number, when multiplied by 4, is equal to 7?

2 What is the smallest possible number of children in the Lee family, if each child has at least one brother and at least one sister?

CREATIVE
Accounting

How to cook
up numbers!

3 A homemaker has 6 lamb chops, each requiring 10 minutes to barbecue on each side. If she has only one grill, which has space for two chops at a time, what is the shortest time she will take before all six lamb chops are barbecued on both sides?

4 In a 100-meter race, Hardy beat Littlewood by 10 m. The two boys plan to have another race, this time Hardy starting 10 m behind the start line. If both boys run at the same rate as before, who will win the race?

Speed Maths
in Practice

11111 x 11111 = 123454321

⑤ A cubic container measuring 12 cm by 12 cm by 12 cm is full of mercury. Ten steel spheres each with a diameter 2 cm is dropped into the container at a time. How many spheres will be completely submerged by the mercury?

(Volume of a sphere = $\frac{4}{3}\pi r^3$)

* ⑥ How many times each day do the hour and minute hands form an angle of 90 degrees between them?

7 How many pancakes measuring 4 cm wide can be cut from a larger pancake of diameter 12 cm?

8 To print a mathematics book, 252 digits were needed. How many pages are there in the book?

Nøthing
but
ZERØ

K C Yan

2011

* **9** Robert is selling 3 computers for $8400 or 5 computers for $13,400. Either transaction yields the same profit. If Robert sold 22 computers, how much did he earn?

10 The sum of all the digits from 1 to 12 is

$$1 + 2 + 3 + 4 + 5 + 6 + 7 + 8 + 9 + (1 + 0) + (1 + 1) + (1 + 2) = 51.$$

Find the sum of all digits from 1 to 999,999.

An order of magnitude is a range of which the upper limit is 10 times as large as the lower limit.

Bravo Singapore for Being the Top Ten

Worst Top Ten for the Merlion City

10. Highest number of gamblers per thousand population.

9. Highest number of laws per square kilometre.

8. Highest number of complainers per household.

7. Highest number of idol-worshippers per square kilometre.

6. Highest expenditure on tuition fees per household.

5. Highest number of *kiasuists* per household.

4. Highest number of respiratory cases among developed nations.

3. Highest number of spectacled citizens per hundred population.

2. Highest number of unreported incest cases among newly-developed nations.

1. Highest number of child-gluttons per hundred families.

1. What was the first day of the 20th century?

2. The product of two whole numbers is 10,000,000. None of them is a multiple of ten. What is the sum of the two numbers?

3. How many years elapsed from the rise of the Roman Empire in 753 B.C. to the fall of the Roman Empire in 422 A.D.?

4. The area of the circle is 18π cm^2.
What is the area of the square?

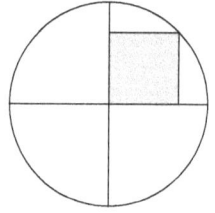

*5. A wooden cube measuring 1 m on each side is cut into many millimeter cubes. If all these smaller cubes were to be placed on top of each other, what would be the height, in kilometers?

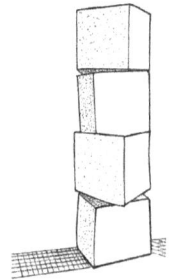

6 Curtis can fire 3 shots in 3 seconds while Bob can fire 5 shots in 5 seconds. Assuming that timing starts when the first shot is fired and ends with the last shot, and the shots themselves take no time, who can fire 8 shots in a shorter time?

7 Susan has two children. They are not both boys. What is the chance that both children are girls?

White + White = Black

* ⑧ Dr. Yan drives along a long stretch of road on which a regular service is in operation. He is driving at a constant speed. He observes that every 3 minutes he meets a train and every 6 minutes a train overtakes him. How often does a train leave the terminal station at one end of the route?

⑨ At a math meeting, 14 members like algebra or geometry, 8 like algebra only, and 5 like both algebra and geometry. How many like geometry only?

* ⑩ Hamid always arrives at the station at exactly 6 o'clock to pick up his wife Siti and drive her home. One day, Siti arrives an hour early, starts walking home, and is eventually picked up. She arrives home 12 minutes earlier than usual. How long did Siti walk before she met her husband?

Dress up mathematically!

26 Mathematical Quickies & Trickies

Lightning Calculators

Lightning calculators demonstrate that the human brain is capable of feats that remain largely unexplained. Most of these idiots-savants demonstrated their extraordinary gift in early childhood. The lightning calculator remains an unexplained phenomenon.

Characteristics of lightning calculators:

• Remarkable memories, often in areas other than computation.

• Rapid recall, a love of arithmetical computations and arithmetical shortcuts, mathematical precocity and a good visual imagination.

A combination of these factors in a human being will produce a lightning calculator. Let's look at how fast some lightning calculators performed certain tasks.

Lightning calculator	Task	Time
Thomas Fuller (1710-1790) African slave	How many seconds are there in a year and a half?	Two minutes.
	How many seconds has a man lived who is 70 years, 17 days and 12 hours old?	One and half minutes.
	A farmer has 6 sows, and each sow has 6 female pigs the first year, and they all increase in the same proportion, to the end of 8 years, how many sows will the farmer then have? ($7^8 \times 6 = 34\ 588\ 806$)	Ten minutes.
Jedediah Buxton An illiterate English farmer (1707-1772)	How many barley corns, vetches, peas, beans, lentils and grains of wheat, oats and rye would fill a space of 202 680 000 360 cubic miles? And also how many hairs laid side by side long (and taking 48 hairs laid side by side to measure one inch across) would fill the same space?	One month.

Lightning calculator	Task	Time
Zacharias Dase 19th century travelling genius (1824-1861)	Multiplying 79 532 853 by 93 758 479. Multiplying two 20-digit numbers. Multiplying two 40-digit numbers. Multiplying two 100-digit numbers. (Without writing anything down)	54 seconds. 6 minutes. 40 minutes. 8 hours 45 minutes.
Vito Mangiamele 19th century, the son of a Sicilian shepherd.	What is the cubic root of 3 796 416? What is the 10th root of 282 475 249? (7)	Half a minute.
Zerah Colburn American's farmer boy (1804-1839) Died at 35.	What is the square root of 106 929? (327) when asked at eight years old. What is the cube root of 268 336 125? (645) Was able to tell whether a number is prime or not. If not prime, he would give its factors. How many days and hours since the Christian Era commenced, 1811 years? (661 015 days, 15 864 360 hours)	Almost readily. Almost instantly. 20 seconds.
George Parker Bidder (1806-1878) The Calculating Engineer	What is the square root of 119 550 669 121? (345 761)	30 seconds.
Shakuntala Devi (b. 1940-) Uneducated daughter of poor parents.	Find the value of (25 842 + 111 201 721 + 370 247 830 + 55 511 315) × 9878. (5 559 369 456 432) Amasterofthecriss-crossmethod,Deviheldthe record of the fastest timed multiplication of two 13-digit numbers on paper. Her feat is recorded in the *Guinness Book of World Records* as an example of a "Human Computer."	Less than 20 seconds.

1 Without expressing the numbers into prime factors, find the highest common factor (or greatest common divisor) of 943 and 1357.

2 Without repeated cancellations, reduce $\dfrac{370\ 368}{617\ 280}$ to lowest terms.

Recall your divisibility tests!

3 Solve the pair of equations: $4567x + 5433y = 14\ 567$
$$5433x + 4567y = 15\ 433$$

4 The sum of two numbers is 56. The product of the numbers is 14. Find the sum of the reciprocals of the numbers.

The reciprocal of 2 is $\frac{1}{2}$.

*   5   At a hotel casino, gamblers can only use $5 and $7 chips. What is the largest bet that cannot be placed using the chips?

6   A smaller pipe can fill a tank in 25 min. A bigger pipe can fill the same tank in 15 min. How long will it take for both pipes to fill the tank together?

* **7** Mr. Tan sent a courier from Singapore to China to reach his uncle in a week. His uncle sent a courier from China to Singapore in 8 days. The distance was 4000 km. In how many days will they meet?

8 A van takes an extra 40 min to cover a distance at 60 km/h than at 80 km/h. Find the distance covered.

9 The World Jumbo Lift can hold either 120 adults or 144 children. Given that 90 adults are already in the lift, how many children can still be admitted?

10 There are 45 poles on opposite sides of *Math Street*. The distance between two adjacent poles is 25 m. The poles on one side are arranged such that each pole fills the gap between two other poles on the opposite side. How long is *Math Street*?

27 Mathematical Quickies & Trickies

Geometrical Quickies 4

Example 1

Given the unit square, the interior shaded square is formed by joining each vertex of the unit square to the midpoint of its nonadjacent side. Find the shaded area.

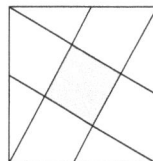

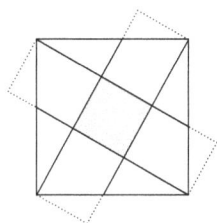

Solution:

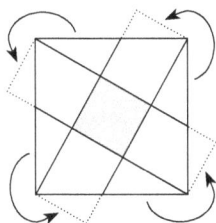

Shaded area = $\frac{1}{5}$ × area of square

$= \frac{1}{5}$ square unit

Example 2

Find the area of the trapezoid, as shown in the figure.

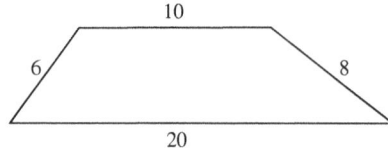

Solution:

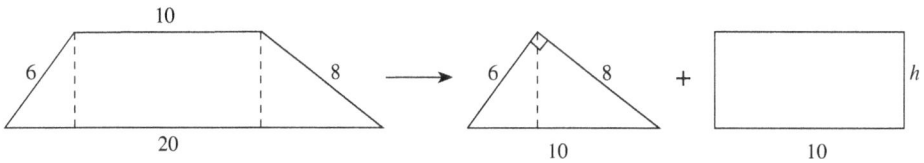

$10 \times h = 6 \times 8$

$h = 4.8$

Area of trapezoid $= \dfrac{1}{2} \times 6 \times 8 + 10 \times 4.8$

$= (24 + 48)$ square units

$= 72$ square units

> We can also find the area by first joining two similar trapezoids to form a rectangle.

168

1. A ship leaves on a long journey. When she is 180 km from the shore, a plane, whose speed is 10 times that of the ship, is sent to deliver mail. How far from the shore does the plane catch up with the ship?

2. If half of 5 were 3, what is one-third of 10?

3. Ten football hooligans arrive at their motel, all drunk. Their room keys are all mixed up. There is no identification number on the keys, and the rooms are all locked. What is the most number of trials required to sort out all the keys?

4 If 70% of the political dissidents have lost a finger, 75% one toe, 80% an ear, and 85% one arm, what percentage must have lost at least all four?

Looks like being in Opposition may be dangerous to your life!

5 Joe went jogging, first on a level road, then up a hill. He then turned around and came back the way he went. He ran at a rate of 8 km/h on the level, 6 km/h going uphill, and 12 km/h going downhill. If Joe spent 2 hours jogging, how far did he go?

6 What the value of $\sqrt{2 + \sqrt{2 + \sqrt{2 + \sqrt{2 + \ldots}}}}$?

7 Find three digits such that the product of one pair is 8, and the product of the other pair is 9.

8 (a) At a math meeting, each of the 15 members shook hands with each of the other members. How many handshakes took place?

(b) If each of the 15 members sat at a round table and hugged with the members to his or her immediate left and immediate right, how many hugs took place?

* 9 A train moving at 70 km/h meets and is passed by a train moving at 50 km/h. A passenger in the first train observes that the second train takes 4 seconds to pass her. How long is the second train?

10 Mother bought some sugar for $13.60. Had the sugar cost 12 cents a kilogram less, she would have received 3 kilograms more for the same amount. How many kilograms did she buy?

A FLYING QUICKIE

The Hungarian mathematician, John von Newmann (1903-1957) was reported to have worked out a series of decreasing terms and computed the sum in solving the following quickie.

> Two trains, A and B, start out 140 miles apart, heading toward each other, A at 40 miles per hour and B at 30 miles per hour. At the same time, a fly that travels at 65 miles per hour leaves A, flies to B, returns to A again, and continues crossing in this manner. How far will the fly have flown by the time they pass?

Can you outsmart Dr. von Newmann by using a tricky method to solve the quickie?

Answer: 130 miles

28 Mathematical Quickies & Trickies

Some Calculator Quickies

Without using algebra, how would you solve these questions? Calculators may be used.

(a) The product of two consecutive whole numbers is 56,406.
     What are the two numbers?

(b) The product of three consecutive odd numbers is 357,627. Find them.

(c) The sum of two consecutive square numbers is 1405. What are they?

(d) The volume of a cube is 200 cm^3. Find the length of the edge of the cube as accurately as you can with a calculator.

(e) What is the remainder when 89,328 is divided by 729?

(f) Find a way of using the $\boxed{\sqrt{}}$ key to evaluate $\sqrt[3]{200}$.

(g) What is the smallest number x which gives the answer 0 to the calculation $1/x$ on your calculator?

① Using the numbers 1, 2, 3, 4, 5 once each in their natural order, how many ways can you form the number 15?

For example, $1 + 2 + 3 + 4 + 5 = 15$

$(1 + 2 \times 3 - 4) \times 5 = 15$

② Using the numbers 1, 2, 3, 4 once each in their natural order and +, −, ×, ÷, √, and by forming powers, can you get every number from 0 to 29? You are not allowed to put 1, 2, 3, 4 together to make numbers like 123 or 34.

3 On the 1st day, put one cent in the piggy bank.
On the 2nd day, put two cents in the piggy bank.
On the 3rd day, put four cents in the piggy bank.

If one kept doubling the amount every day,
for a month, how much would one save?

4 A manufacturer delivered 320 sets of calculators to a shop. It charged 70 cents for every set delivered but it had to pay $7.30 for every broken calculator. If the shop owner paid a total of $184.00 for the delivery, how many calculators were broken?

5 *Curious Division*

Consider the following:

$$\frac{1111111111111111}{9} = 1234567901245679$$

Why is there no figure 8 in the answer?

6 *A Call Sign for the Police*

Multiply 37037037 by 27.

By what number should 37037037 be multiplied to get the same result upside down?

7 Work out the following:

$$11 \times 91 = \underline{\hspace{2cm}}$$
$$101 \times 9901 = \underline{\hspace{2cm}}$$
$$1001 \times 999\ 001 = \underline{\hspace{2cm}}$$

What is the next product? And what is its answer?

Use any patterns in the results to express 1 000 000 000 000 001 as the product of two whole numbers.

8 Work out the following.

$$3 \times 37 = \underline{\hspace{2cm}}$$
$$33 \times 3367 = \underline{\hspace{2cm}}$$
$$333 \times 333\ 667 = \underline{\hspace{2cm}}$$

What would be the next product and its answer?
Use any patternful result to express 111 111 111 111 111 as the product of two numbers.

(9) Without using a calculator, give an estimate of 2^{83}.

(10) When the numbers 4572301 are entered, the screen of a disabled calculator shows 1542630.

What number will show up if 89 is pressed?

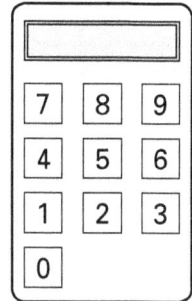

7	8	9
4	5	6
1	2	3
0		

Mathematical Quickies & Trickies

Mathematical Quickies & Trickies 1 (p. 5)

1. Since 133 = 7 × 19, there are 19 employees, each working for 7 hours a day.

2. Esther is 13 years and 6 months old. Ruth is only 6 months old.

   In another 12 years Ruth will be 12 years and 6 months old.

3. Half an hour before they meet, the trains will have travelled $\frac{1}{2} \times 55$ km and $\frac{1}{2} \times 75$ km, respectively. The distance between them will then be

   $\frac{1}{2} \times (55 + 75)$ km = 65 km.

   Notice: The answer does not depend on the distance between the two cities.

4. They played nine games. Solomon won six games.

5. The empty bottle costs 5¢ and the lemonade costs 35¢.

6. There are only three women: the grandmother, the mother, and the daughter.

7. Whatever their speeds, both trains are at the same distance from Samaria.

8. Ten.

   Mr. and Mrs. Pilate, two daughters and their husbands, and four children.

9. 3 m 38 cm or 338 cm.

   The first 3 meters of cloth is shorter by 3 × 4 cm = 12 cm. The next 50 cm of cloth will be correctly measured because the error is only after 92 cm. Thus the actual length of the cloth will be (350 − 12) cm = 338 cm.

10. Only two hours.

    The alarm would go off at 11 o'clock at night.

Mathematical Quickies & Trickies 2 (p. 12)

1. Three.

   If you pick one of each color, the third must match one of them.

2. Eleven, all except February. A 31-day month still contains 30 days.

3. 9^{9^9}.

   9^9 is 387,420,489. That number to the ninth power is a 77-digit number.

4. 63 games were played. Each game eliminates a player.

   After 64 players have played, only one player remains unbeaten.

5. The chance is one in six. *Remember*: Dice don't have memories.

6. Give the fifth girl her apple in the basket.

7. 2 bicycles and 5 tricycles.

   Suppose all the items were bicycles. Then 14 wheels would be used.

   That leaves us with 5 wheels, which must come from 5 tricycles.

   Therefore, there are 5 tricycles and 2 bicycles.

   Check: 5 × 3 + 2 × 2 = 15 + 4 = 19

8. Pages 35 and 36.

9. 66 seconds.

   When the clock struck six, there were only 5 intervals between strokes, and each interval was 30 ÷ 5 = 6 seconds.

   Between the first and twelfth strokes there will be 11 intervals of 6 seconds each; therefore, 12 strokes will take 66 seconds.

10. $2.

    Using algebra,   Using a model

    $x = 1 + \frac{x}{2}$

    Solve x.

    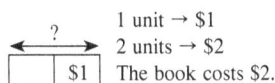

    1 unit → $1
    2 units → $2
    The book costs $2.

1. A pound of metal is always worth more than half a pound of the same metal.

2. The difference in age is still 30 years, so I must be 30 if my father is twice as old.

3. Four boys and three girls.

4. $1\frac{1}{2}$.

   $\frac{1}{2} = \frac{1}{3}x$.

5. 15 minutes.

   In $3\frac{1}{2}$ hours the alarm clock has become 14 minutes slow. After another 14 minutes the alarm clock will have lost approximately one more minute. Its hands will show noon in 15 minutes.

6. Not 8 km/h, but 6 km/h.

   If x represents half the distance, then average

   speed = $\dfrac{2x}{\frac{x}{12} + \frac{x}{4}}$.

   Can you solve the question using a non-algebraic method?

7. $166\frac{2}{3}$ m.

   The speed of the passenger in the first train, in relation to the movement of the second

   train, is $55 + 45 = 100$ km/h, or $\dfrac{100 \times 1000}{60 \times 60}$

   $= \dfrac{1000}{36}$ m/s.

   Therefore, the length of the second train is

   $6 \times \dfrac{1000}{36} = 166\frac{2}{3}$ m.

8. 4 times.

   He walked one third of the way, or half as far as he cycled, but it took him twice as long. Therefore, he cycled four times as fast as he walked.

9. 23 cents.

   If Joe had 2 or more cents, then he and Jane would have enough money to buy a pen, since Jane was only 2 cents short. So Joe had 1 cent or 0 cents. Since Joe had something (1 cent), the price of the pen must then be 23 cents.

10. 500 kg.

    Note: The dry mass of the grapes remains unchanged. For the fresh grapes it amounts to 1% of the whole weight, or 10 kg.

    One week later, the grapes had only 98% water. The 10 kg would now represent 2% of the total mass.

    Thus, the total mass was then 500 kg.

    This means that over 50% of the water had evaporated from the grapes one week later.

1. 5 corners.

2. A round cover cannot fall in — almost any other shape can.

3. Not 3, but 4 seconds.

   1 interval → 2s

   2 intervals → 4s

4. One dog.

5. There will be as much oil in the water flask as there will be water in the oil flask. *Do you know why?*

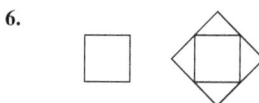

6.

7. The beggar was a woman.

8. 90 km/h is wrong.

   Once the man has traveled for half the journey at 30 km/h, it is impossible to average 60 km/h for the total journey.

   Assume that the distance from A to B was 30 km, so averaging 30 km/h for the first half of the journey would use up 1 hour. But the total journey is 60 km. The average 60 km/h for the total journey means that the man needed to cover the total journey in 1 hour. He had used up his 1 hour in just traveling half the distance. Therefore he cannot average 60 km/h for the total journey.

9. Holes.

10. Set both 7-min and 11-min timers on. When the 7-min timer is up, start boiling the egg for another 4 minutes, the time for the remaining sand to flow down the 11-min timer. Then turn the 11-min timer upside down for another 11 more minutes. Thus the egg will be boiling for a total of 4 + 11 = 15 minutes.

Mathematical Quickies & Trickies 5 (p. 34)

1. 10,000.

2. $(5 - 1) \times 3 \times 3$ or $(5 + 1) \times (3 + 3)$.

3. $89 \times 7349 - 89 \times 7348 = 89 \times (7349 - 7348)$
$$= 89 \times 1$$
$$= 89$$

   Ann should add 89 to get the correct answer.

4. $14 \times 25 \times 36 = 12,600$

5. 0.8571429.

   They are the decimal equivalents of the following fractions: $\frac{1}{3}, \frac{2}{3}, \frac{3}{4}, \frac{4}{5}, \frac{5}{6}, \frac{6}{7}$

6. He is probably a shopkeeper or some merchant.

   $12 + 25\% = 16$ is correct because it means $12 + (25\%$ of $16) = 16$.

   The 25% addition represents the profit margin on the 12 he wishes to add. If $15 was the selling price, the extra $3 would only give a 20% profit.

7. $28 \times 9998 - 28 \times 9978 = 28 \times (9998 - 9978)$
$$= 28 \times 20$$
$$= 560$$

   Paul should add 560.

8. 740938271.

   Only the number 0 is correctly produced. When one of the other numbers is entered, the two accompanying numbers in the same column appear on the display. For instance, if you enter the number 6, then 93 will appear on the display.

9. Between the 7 and the 2.

   The order depends on the number of 'segments' required to make up the digital number.
   The number 7 uses three segments, 2 uses five, and 4 uses four.

10. $-33 - 3 - (3 \div 3)$.

11. 3, 24.

Mathematical Quickies & Trickies 6 (p. 40)

1. Nine rectangles.

2. (a) 11,231.
   Two methods are:
   $$1235 + 9996 = 1235 + (10,000 - 4)$$
   $$= 11,235 - 4$$
   $$= 11,231$$
   $$1235 + 9996 = 1231 + 4 + 9996$$
   $$= 1231 + 10,000$$
   $$= 11,231$$

   (b) 8999.
   $$9997 - 998 = (10,000 - 3) - (1000 - 2)$$
   $$= (10,000 - 1000) - 3 + 2$$
   $$= 9000 - 1$$
   $$= 8999$$
   $$9997 - 998 = (9777 + 3) - (998 + 2)$$
   $$= (10,000 - 1000) - 3 + 2$$
   $$= 9000 - 1$$
   $$= 8999$$

3. Ah Koon was born on February 29.
   The likelihood that a man will be more than n years old on his nth birthday is as little as one to 1460 or slightly better if we allow for seasonal trends.

4. 5 six-meter and 9 seven-meter pipes.
   If all 14 pipes were each 6 m long, then that would cover a distance of $(14 \times 6) = 84$ m. The remaining $(93 - 84) = 9$ m must have come from $9 \div (7 - 6) = 9$ pipes each of 7 m long.
   Thus there are 9 seven-meter pipes and $(14 - 9) = 5$ six-meter pipes.

5. $60.
   Using a non-algebraic method, the number of 5-dollar notes is $\frac{8 \times 3}{5 - 3} = 12$.
   The book costs $12 \times \$5 = \$60$.

6. 10.10 a.m.

   Both bulbs flash together every 14 minutes. Thus, both bulbs will flash together in $5 \times 14 = 70$ minutes after 9:00 a.m. Hence, both bulbs will flash again at 10:10 a.m.

7. Not 27 but 54 marbles.

8. 1026 km.

Let the car take x hours to travel from A to B.

Then $57x = 38(x + 9)$

$\qquad x = 18$

Distance from A to $B = 57 \times 18$

$\qquad\qquad\qquad\qquad = 1026$ km

9. 1:20 p.m.

$E + \dfrac{4}{5}E = 24$

$E = \dfrac{40}{3}$ h $= 13$ h 20 min

10. $13\dfrac{1}{3}$ km/h.

Mathematical Quickies & Trickies 7 (p. 46)

1. Not 8 cm, but 4 cm.

2. The two chess players were not playing each other.

3. Three strokes.

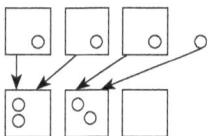

4. 5 units.

5. The sequence is based on the first letter of the number spelled, in reverse alphabetical order:

Zero, Two, Three, Six, Seven, One, Nine, Four, Five, Eight.

6. Six cigarettes.

He can assemble 5 cigarettes and smoke them. Then he will have 5 butts left over, from which he can assemble a 6th cigarette.

7. 3 boxes and 4 marbles.

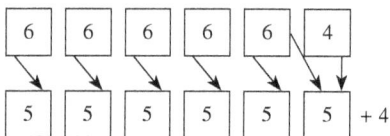

8. 23 chickens and 12 rabbits.

If all 35 animals each had two feet, that would yield 70 feet.

So, the extra $94 - 70 = 24$ heads must have come from the rabbits, with each rabbit contributing another two feet.

Thus, there are $\dfrac{24}{2} = 12$ rabbits, and

$(35 - 12) = 23$ chickens.

9. If 24 billion hours later, it will be 9 p.m. Four hours before that, it will be 5 p.m.

10. The magnifying glass cannot increase the magnitude of the angle. While the arc measuring the angle increases, its radius increases proportionately. The net effect is that the magnitude of the angle remains unchanged.

Mathematical Quickies & Trickies 8 (p. 52)

1. 27 cuts. (9 + 9 vertical cuts + 9 horizontal).

2. 6 cells and 34 prisoners.

6	6	6	6	6	4
5	5	5	5	5	5	+ 4

3. Pages 145 and 146 are opposite to each other.

4. 28 hours.

After 27 hours, the frog is at the 27-foot mark. One hour later, it is on top of the well and ready to get out.

5. 128 poles ($= 4 \times 31 + 4$).

6. $22\dfrac{1}{2}°$.

The hour hand moves the same fractional distance between two and three (30°) as the minute hand has moved of a complete rotation $\left(\dfrac{15}{60} = \dfrac{1}{4}\right)$.

The angle is $30° - \dfrac{1}{4} \times 30° = 22\dfrac{1}{2}°$.

7. One pound of feathers.

The two "pounds" are not the same. A pound of gold is measured in troy weight (5760 grains), while a pound of feathers is measured by avoirdupois weight (7000 grains).

8. Once properly set, the clock that loses a minute a day will have to lose 12 hours, or 720 minutes before it is right again. And if it loses only a minute a day, it will take 720 days to lose 720 minutes. In other words, it is correct only once about every two years. But the clock that does not run at all is correct twice a day!

9. Seven 5 s.

The numbers that have 5 as factor are:

5, 10, 15, 20, 25, 30.

Since $25 = 5 \times 5$, these number will yield seven 5 s.

10. 2.72 h = 2 h + (60×0.72) min, i.e., 2 h 43 min.

Einstein will arrive at 4:43 p.m.

Mathematical Quickies & Trickies 9 (p. 59)

1. 400,006.

$499,992 - 99,986$

$= (500,000 - 8) - (100,000 - 14)$

$= (599,000 - 100,000) + (14 - 8)$

$= 400,000 + 6$

$= 400,006$

2. The next four numbers are: 1, 4, 1, 5, being the number of times a clock that chimes on the hour and the half hour chimes.

3. Since one of the terms in this series will be $(x - x)$, which equals zero, the product of the entire series is zero.

4. Mr. Kiasu has to drive at an infinite speed to average 100 km/h for the race—he must complete the whole 1000 km in 10 hours to attain the required speed. That is impossible because he has already used up his 10 hours to cover the first half of the race. He will have to finish the race in zero time!

5. Seven problems.

$35 = 5 \times 7$

$50 = 2 \times 5 \times 5$

LCM $(35, 50) = 2 \times 5 \times 5 \times 7 = 350$

Number of correct answers $= \dfrac{350}{2 \times 5 \times 5} = 7$

Number of incorrect answers $= \dfrac{350}{5 \times 7} = 10$

6. \$92.

12 months $\rightarrow$ \$100 + H

7 months $\rightarrow$ \$20 + H

Find the LCM of 7 and 12.

Show that \$700 + 7H = \$240 + 12H.

7. Paul because $\left(\dfrac{x}{30} + \dfrac{x}{40}\right)$ is greater than $\dfrac{2x}{35}$.

8. 160 km/h.

Since the driver traveled 3 km at 140 km/h, it took him $\dfrac{60}{140} \times 3 = \dfrac{9}{7}$ min to cover the distance.

Similarly, 1.5 km at 168 km/h would take him $\dfrac{60}{210} \times \dfrac{3}{2} = \dfrac{15}{28}$ min, and 1.5 km at 210 km/h would take him $\dfrac{60}{210} \times \dfrac{3}{2} = \dfrac{3}{7}$ min.

Total time taken for the 6 km

$= \left(\dfrac{9}{7} + \dfrac{15}{28} + \dfrac{3}{7}\right) = \dfrac{63}{28}$ min

Average speed $= \dfrac{6 \times 28}{63}$ km/min, or

$\dfrac{6 \times 28 \times 60}{63}$ km/h.

9. $5^{1/3}$ min.

10. Even. If you were to ask everyone in the world how many hands he or she has shaken, the total would be even because each handshake would have been counted twice — once each by the two people who shook hands. A group of numbers whose sum is even cannot contain an odd number of odd numbers.

Mathematical Quickies & Trickies 10 (p. 66)

1. $4 \div 4 + 4 + 4$

2. 14, 15, 16

3. $431 \times 52 = 22,412$

4. $8 \div 17$

5. $888 + 88 + 8 + 8 + 8$

6. Some possible answers are:

$1 + (9 - 9) \div 9 = 1$

$1 + (9 + 9) \div 9 = 3$

$1 \times 9 - (9 \div 9) = 8$

$1 \times (9 + 9) - 9 = 9$

$1 + 9 + 9 - 9 = 10$

$1 + 9 + (9 \div 9) = 11$

$1 + 99 \div 9 = 12$

$19 - (9 \div 9) = 18$

$19 + 9 - 9 = 19$

$19 + (9 \div 9) = 20$

7. The calculator and telephone keypads have different layouts, so when Ian pressed the keys on the telephone the numbers were wrong except for the middle numbers (4, 5, 6).

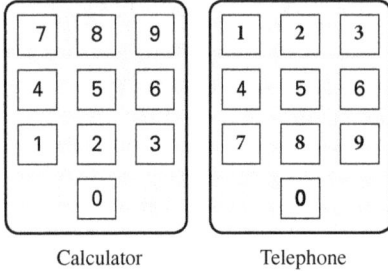

Calculator   Telephone

8. Move one line from the equals to the minus. Then turn the book upside down!
The equation then reads 569 – 288 = 281, a correct sum.

9. To delete digit 2: (43 987 526 – 26 + 60) ÷ 10
= 4 398 756
Or, 43 987 526 – 43 987 520 + 4 398 750
= 4 398 756
To delete digit 3: 4 398 756 – 4 300 000
+ 400 000 = 498 756
To delete digit 4: 498 756 – 400 000 = 98 756 To
delete digit 5: (98 756 – 56 + 60) ÷ 10 = 9876 To
delete digit 6: (9876 – 6) ÷ 10 = 987
To delete digit 7: (987 – 7) ÷ 10 = 98
To delete digit 8: (98 – 8) ÷ 10 = 9

10. 111 111 111 × 111 111 111
= 123456678987654321

11. 61 × 52 × 43 = 136,396

12. Let $A = \dfrac{10^{11} + 1}{10^{10} + 1}$ and $B = \dfrac{10^{12} + 1}{10^{11} + 1}$

$\dfrac{B}{A} = \dfrac{10^{12} + 1}{10^{11} + 1} \div \dfrac{10^{11} + 1}{10^{10} + 1}$

$= \dfrac{(10^{12} + 1)(10^{10} + 1)}{(10^{11} + 1)(10^{11} + 1)}$

$= \dfrac{10^{22} + 10^{12} + 10^{10} + 1}{10^{22} + 2 \times 10^{11} + 1}$

$= \dfrac{10^{22} + 1 + (10^2 + 1) \times 10^{10}}{10^{22} + 1 + 20 \times 10^{10}}$

Thus $B > A$

Hence, $\dfrac{10^{11} + 1}{10^{10} + 1}$ is smaller than $\dfrac{10^{12} + 1}{10^{11} + 1}$.

1. 6 years old.
The calendar number goes from 1 B.C. to 1 A.D. without a zero, unlike the real number line which has a zero. From 3 B.C. to 4 A.D. there are only 6, rather than 7 years, because there is no zero year.

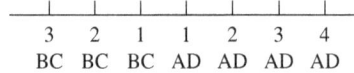

| 3 | 2 | 1 | 1 | 2 | 3 | 4 |
| BC | BC | BC | AD | AD | AD | AD |

2. 21 letters.

3. One.
A *birthday* is the day one was born. One might have celebrated it ten times over ten years, but one still had only one birthday, namely the day one was actually born.

4. Five sheets.
All books begin with page 1 on the right-hand side. Page 2 is on the other side of page 1. Any pair of consecutive pages beginning with an odd number is one sheet of paper. Any pair of consecutive pages beginning with an even number would be 2 sheets.

5. 35 triangles.

6. Three children.

7. 28 pounds of wool.

Sheep	Wool	Days
$1\frac{1}{2}$	$1\frac{1}{2}$	$1\frac{1}{2}$
3	3	$1\frac{1}{2}$
6	6	$1\frac{1}{2}$
6	$6 \times \frac{2}{3}$	1
6	$6 \times \frac{2}{3} \times 7 = 28$	7

8. 23 matches.
There are 23 losers and only one winner.

9. 10 s.

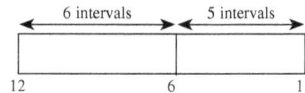

10. 20 (7, 17, 27, 37, 47, 57, 67, 70, 71, 72, 73, 74, 75, 76, 77, 78, 79, 87, 97).

1. 12 stamps.

2. 3 coins.

After the second coin, we have either two balls of the same color, or two balls of different colors. A third coin would deliver a ball that had to match one of the colors.

3. Tangram set: $7.50; magazine: $5.00.

4. 5 minutes.

It is assumed that each fisherman takes 5 minutes to catch a single fish. The time element would never change regardless of how many fishermen and fish are involved.

5. None.

A hole is a cavity with nothing in it.
Therefore, there is nothing to be removed.

6. 4 pounds of $10 silver coins.

A heavier weight would always be worth more than a lighter weight of the same matter (silver). The worth of each individual coin is immaterial because worth is dependent on total weight.

7. There is no difference.

80 min = 1 h 20 min

8. House numbers: $\boxed{1}$ $\boxed{12}$ $\boxed{144}$

9. 3 feet.

Because a tree grows taller from the top, the rest of the tree remains at the same height.

10. 45 cents.

Cut each link in any section to obtain three open links. One of these links will join Sections A and B (say), the second link joins Sections B and C, while the third link joins Sections C and D, and weld all three links. This makes three cuts (15 cents) and three weldings (30 cents).

1. 7 hours.

2. 11 days.

3. 4 cans.

Area of ceiling = (4.5×5.5) m^2

Two coats require an area of $(2 \times 4.5 \times 5.5)$ m^2.

Amount of paint needed = $\dfrac{2 \times 4.5 \times 5.5}{16}$
= 3.1 litres

Hence 4 cans of paint are needed.

4. Once. Then you are subtracting 2 from 15.

5. (i) $\dfrac{1}{12}$ (ii) $\dfrac{1}{3}$

The factors of 60 are 1, 2, 3, 4, 5, 6, 10, 12, 15, 20, 30 and 60 — there are 12 of them.

(a) Fraction of all factors that have a factor

$20 = \dfrac{1}{12}$.

(b) Factors of 60 that are also factors of both 2 and 5 are 10, 20, 30 and 60 — there are 4 of them.

Fraction of all factors that have factors

of both 2 and 5 = $\dfrac{4}{12} = \dfrac{1}{3}$.

6. 24 stamps.

There are 12 in a dozen, and 12 dozens in a gross.

7. $12\frac{1}{2}$ minutes. Boil them all at the same time.

8. Three oranges.

If John were to have $1\frac{1}{2}$ times as many oranges as he actually had, he would have what he actually had plus $1\frac{1}{2}$ oranges. So $1\frac{1}{2}$ oranges represent half of what he had. Therefore, John must have had three oranges.

9. 18 buses.

$640 = 37 \times 17 + 11$

A total of 37×17 students would be on 17 buses, with the remaining 11 students on the 18th bus.

10. $100.

First transaction = $550 – $500 = $50
Second transaction = $600 – $550 = $50
He made a total of $50 + $50 = $100.

1. 12,111.

2. Bobby packs a box in 9 minutes.

   Hint:

   Let Bobby and Betty take x and $(x + 3)$ minutes, respectively.

   Then $\frac{6}{x} + \frac{4}{x + 3} = 1$.

   Try solving the question using a non-algebraic method.

3. 63.25 cubits.

   By Pythagoras' Theorem, they are apart by

   $\sqrt{20^2 + 60^2} \approx 63.25$ cubits

4. 10 ounces must be added.

   Solve $0.26x + 0.06(10) = 0.16(x + 10)$

5. On day 15, the lake is half covered.

6. 31.

   $93{,}093 = 3 \times 7 \times 11 \times 13 \times 31$

7.

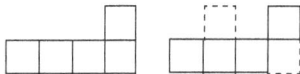

8. 24 minutes.

   Speed of the boy relative to the man
   = 6 km/h

   Speed of the fishermen relative to the man
   = (6 − 5) km/h
   = 1 km/h downstream

   The difference in speed between the boy and the fishermen is 5 km/h.

   Therefore it will take $\frac{1}{5} \times 2$ h = 24 min

   An intuitive solution

   If we approach the situation from the point of view of the boy on the raft, then we can ignore the rate at which the boy and the fishermen are moving relative to the man. [The man (observer) is irrelevant in this situation.] Besides, we can ignore the rate of the current, since it is affecting the boys and the fishermen equally.

   We are left with the rate at which the fishermen are paddling: 5 km/h.

   Unlike the first method that involves three steps, this method allows us to find the rate in one step.

9. If the first person was a liar, he would lie about himself and say that he was a truth-teller.

   If the first person was a truth-teller, he would tell the truth about himself and say that he was a truth-teller.

   Whether the first man was a truth-teller or a liar, he would say that he was a truth-teller.

   The second man must have been a truth-teller. So was the first man because the second man said so.

10. 31 cents.

    Note that $3193 = 31 \times 103$
    (31 is a factor of 3193, and 103 is prime.)

1. $2\frac{1}{2}$ times higher.

2. The decimal 7.8.

3. $1 + 2 + 3 = 1 \times 2 \times 3$.

4. 4523.

   If $ABCD = 1234$, then $DCBA = 4321$. It is impossible to arrive at a sum as large as 12,300. If $ABCD = 3456$, then $DCBA = 6543$. It is impossible to arrive at a sum as small as 12,300. Therefore $ABCD = 2345$. Then $DCBA = 5432$. Hence the four Xs stand for 4523.

5. 3,265,920.

   The ten digits can form $10 \times 9 \times 8 \times 7 \times \cdots \times 3 \times 2 \times 1 = 3{,}628{,}800$ different 10-digit whole numbers. Since the 10-digit number cannot start with zero, the number of different 10-digit numbers is, therefore, given by

   $$3\ 628\ 800 - \left(\frac{3\ 628\ 800}{10} \right),$$

   i.e., $3{,}628{,}800 - 362{,}880 = 3{,}265{,}920$.

6. $33\frac{1}{3}$ cups.

   3% of caffeine can be obtained from 1 cup of coffee.

   100% of caffeine can be obtained from

   $\frac{1}{3} \times 100 = 33\frac{1}{3}$ cups of coffee.

7. 14 squares.

   Nine 1×1 squares + four 2×2 squares + one 3×3 square.

8. March 23.

March has 31 days. To have five Fridays, they would fall on the following dates:

1	2	3
8	9	10
15	16	17
22	23	24
29	30	31

Since three of the dates must fall on even-numbered dates, the Fridays would fall on 2nd, 9th, 16th, 23rd, and 30th.

The 4th Friday would be March 23.

9. 76.

If we assign number values to each letter in the alphabet, such as A = 1, B = 2, C = 3, and so on, then

P	L	U	S	
16	12	21	19	= 68

M	I	N	U	S	
13	9	14	21	19	= 76

10. Neither.

5% of seven billion $= \dfrac{5}{100} \times 7 \times 10^9$

$\qquad\qquad\qquad\qquad = 35 \times 10^7$

7% of five billion $= \dfrac{7}{100} \times 5 \times 10^9$

$\qquad\qquad\qquad\qquad = 35 \times 10^7$

Thus, 5% of seven billion $= 7\%$ of five billion

Mathematical Quickies & Trickies 16 (p. 100)

1. The barber makes twice as much money.

2. $\sqrt{2}$ cm.

If one side is the base and the other side is allowed to rotate, the area of the triangle will be greatest when the perpendicular height (or altitude) is maximum.

By Pythagorean theorem, the third side is

$\sqrt{1^2 + 1^2} = \sqrt{2}$.

3. Any two whole numbers (or positive integers), x and y, satisfy the relationship:

$$\text{GCD}\,(x, y) \times \text{LCM}\,(x, y) = xy.$$

4. 132 dates.

Each month has 11 ambiguous dates (a date such as 7/7/98 is not ambiguous), making a total of $(11 \times 12) = 132$ ambiguous dates in a year.

5. 4 square units.

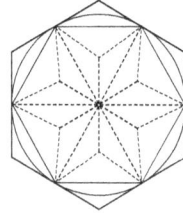

Turn the position of the hexagon.
The lines divide the larger hexagon into 24 congruent triangles, 18 of which form the smaller hexagon.

Since the ratio of the areas is $18 : 24 = 3 : 4$, and the area of the smaller hexagon is 3 square units, the larger hexagon has an area of 4 square units.

6. Ian was born in 1980. He will be 45 years old in $45^2 = 2025$.

7. 90.

There are a total of $(99 - 10 + 1) = 90$ numbers from 10 to 99.

8. 25 problems.

Let x represent the number of correctly solved problems.
Then $50x = 30 \times (40 - x)$
On solving, $\qquad x = 15$
$\qquad\qquad 40 - x = 25$

Alternatively,

Every 5 incorrect solutions nullifies 3 correct solutions.

Thus there are $\dfrac{40}{(5 + 3)} \times 5 = 25$ incorrect answers.

9. One-third.

$1 - \dfrac{3}{4} = \dfrac{1}{4}$

$\dfrac{\frac{1}{4}}{\frac{3}{4}} = \dfrac{1}{3}$

$\dfrac{4}{4}$ exceeds $\dfrac{3}{4}$ by 1 unit, which is $\dfrac{1}{3}$.

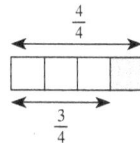

10. 19 years old, since $1 + 2 + 3 + \cdots + 19 = 190$.

Alternatively,

Algebraically, $1 + 2 + 3 + \ldots + n = \dfrac{n(n + 1)}{2}$

$$\frac{n(n + 1)}{2} = 190$$
$$n(n + 1) = 380$$

Since $19 \times 20 = 380$, therefore, $n = 19$.

Mathematical Quickies & Trickies 17 (p. 105)

1. 199,984.

$99,996 + 99,988$
$= (100,000 - 4) + (100,000 - 12)$
$= (100,000 + 100,000) - 4 - 12$
$= 200,000 - 16$
$= 199,984$

2. 0.

$1 + 2 + 3 + \ldots + 3 + 2 + 1 = 81$, which has a remainder 0 when divided by 9.

Or, use "casting out the nines": $1 + 8, 2 + 7, \ldots$— *see Quickies & Trickies 9.*

3. $a^2 + b^2$.

Note that $(a - b)^2 > 0$ for real $a \neq b$.
But $(a - b)^2 = a^2 + b^2 - 2ab > 0$
Therefore, $a^2 + b^2 > 2ab$.

Or, use concrete numbers to show the result.

4. $\dfrac{3}{4}$.

Whether one throws one die twice, or two dice once, the chance is the same. In either case, the throw of one die does not affect the throw of any other die, or the throw of the same die again. There is a 50% chance that one will not get 4, 5, or 6 with the first throw. Similarly, the chance still remains 50% that none of these numbers will come out on the second throw.

Therefore, the chance that none of these numbers will come out on either throw is $\frac{1}{2} \times \frac{1}{2} = \frac{1}{4}$. Hence the chance that one you will get at least one of the numbers, on at least one throw, is $1 - \frac{1}{4} = \frac{3}{4}$.

5. 25,201.

Let N be such smallest number.
Then $(N - 1)$ must be divisible by any of the numbers from 2 to 10 inclusive.
LCM $(2, 3, 4, \ldots, 10) = 2^3 \times 3^2 \times 5 \times 7 = 2520$
Thus N must be of the form $2520n + 1$.

Now $\dfrac{2520n + 1}{11} = 229n + \dfrac{n + 1}{11}$

Since N is divisible by 11, $n + 1$ must be a multiple of 11.

Clearly, $n = 10$ is the smallest integer n satisfying this condition.

Therefore, $N = 2520 \times 10 + 1$
$= 25,201$

6. No.

Profit on the first book $= \$\left(\dfrac{18}{150} \times 50\right) = \$6$

Loss on the second book $= \$\left(\dfrac{18}{72} \times 28\right) = \$7$

He made a $\$7 - \$6 = \$1$ net loss.

7. 1156.

Since 3, 5, 7, and 11 are all prime, the smallest possible positive such number is

$$(3 \times 5 \times 7 \times 11) + 1 = 1156.$$

8. 24 zeros.

Numbers with one factor of 5: 5, 10, 15, 20, 30, 35, 40, 45, 55, 60, 65, 70, 80, 85, 90, 95	Numbers with two factors of 5: 25, 50, 75, 100

Number of zeros = number of factors of 5
$= 16 + (2 \times 4)$
$= 24$

Alternatively
5, 15, 25, 35, …, 95 yield 12 zeros.
10, 20, 30, 40, …, 100 yield 12 zeros.
Number of zeros = 12 + 12 = 24

9. 2, 599, and 601.

Observe that there exists only one even prime, 2. Since the sum of two prime numbers is also prime, one of the two smaller numbers must be 2.

If the other prime is x, then
$x + (x + 2) = 1200$
$2x = 1200 - 2$
$x = 599$

Thus, the prime numbers are 599 and 601.

10. 13 cars.

Observe that the number of cars behind the ex-champion is the same as the number of cars in front of him. All the other cars are both behind and before him.

Let x be the number of cars, excluding that of the ex-champion.

Then $\frac{1}{4}x + \frac{5}{6}x = x + 1$

$\frac{13}{12}x = x + 1$

$x = 12$

Thus, there are 13 cars running in the Grand Prix.

Mathematical Quickies & Trickies 18 (p. 112)

1. (a) 17,000   (b) 96,000

  (a) $136 \times 125 = 136 \times (1000 \div 8)$
                       $= 136 \div 8 \times 1000$
                       $= 17 \times 1000$
                       $= 17,000$

  (b) $256 \times 375 = 256 \times (3000 \div 8)$
                       $= 256 \div 8 \times 3000$
                       $= 32 \times 3000$
                       $= 96,000$

2. 7:00 a.m.
For every real hour, the watch shows only 58 minutes.
Real: 4:00 p.m.   5:00 p.m.   6:00 p.m. ...
           6:00 a.m.   7:00 a.m.
Watch: 4:00 p.m.   4:58 p.m.   5:56 p.m. ...
           5:32 a.m.   6:30 a.m.
The correct time is 7:00 a.m.

3. 132.
$1 \times 2 = 2$, $3 \times 4 = 12$, $5 \times 6 = 30$,
$7 \times 8 = 56$, $9 \times 10 = 90$, $11 \times 12 = 132$.

4. 17 rungs.
Let's assign the middle rung to 0.

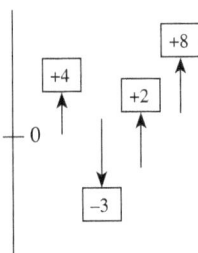

The fireman went up to number 4, down seven rungs to number 3 below 0, then up five rungs to number 2 above.

Finally, he climbed six more rungs to the top.

Thus the top rung must have been number 8 above the middle. Hence the ladder had seventeen $(8 + 8 + 1)$ rungs.

5. $\frac{1}{8}$.
The chance of getting at least one 4, 5, or 6 each time is $\frac{1}{2}$.
Thus the chance of getting 4, 5, or 6 on the first and the second and third throw is $\frac{1}{2} \times \frac{1}{2} \times \frac{1}{2} = \frac{1}{8}$.

6. $9000, $6000, and $2000.
Observe that $\frac{1}{2} + \frac{1}{3} + \frac{1}{9} \neq 1$.
The possessions are to be shared in the proportions: $\frac{1}{2}, \frac{1}{3}, \frac{1}{9}$.
These fractions are equivalent to $\frac{9}{18}, \frac{6}{18}, \frac{2}{18}$, respectively.
Thus $17,000 divided in the proportions of 9, 6 and 2 yields $9000, $6000, and $2000, respectively.

7. 35 books.
Since 20% of 5 is 1 and the number of recreational books must be a whole number, Mr. Yan's total number of books must be a multiple of 5.
Since one-seventh of his books were gifts from friends, the total number of books must also be a multiple of 7.
The first two numbers that are multiples of 5 and 7 are 35 and 70.
Since Mr. Yan has yet to have 50 books, therefore, he must presently have 35 books.

8. 300 calculators.
Suppose of those who had two calculators, each gave one to those who did not have. This means all the mathletes each would then have a calculator. As there are 300 mathletes, there must be also 300 calculators.

9. Three minutes.
In still water, his speed is twice the speed of the current.
If he swims with the current, it is three times the speed of the current.
If he swims against the current, his speed is equal to that of the current in an opposite direction. Let the time taken to swim from one end of the bank to the other end when swimming against the current be t min.
Then the time taken to travel the same distance with the current, and in still water would be $\frac{t}{3}$ min and $\frac{t}{2}$ min, respectively.

191

Therefore, swimming to one end of the bank and back would take t min in still water and $\frac{4t}{3}$ min when there is a current, since it takes t min for the part of the swim that is against the current, and $\frac{t}{3}$ min for the part of the swim that is with the current.

This is equivalent to $\frac{1}{3}$ as long again as the time taken to do the round trip in still water. Since it takes 4 min when there is a current, it must take 3 min when there is no current.

10. 55 deer.

Since it takes 10 deer 10 minutes to jump over a fence, the time interval between jumps is $\frac{10}{9}$ minutes. Therefore, in an hour, there are $\left(60 \div \frac{10}{9}\right) = 54$ such intervals.

Hence 55 deer would jump the fence in one hour.

Mathematical Quickies & Trickies 19 (p. 119)

1. (a) 500,500.

$1 + 2 + 3 + \ldots + 999 + 1000$
$= (1 + 999) + (2 + 998) + (3 + 997) + \ldots +$
$\quad (499 + 501) + 500 + 1000$
$= (499 \times 1000) + 1500$
$= 500,500$

(b) 49,995,000.
$1 + 2 + 3 + \ldots + 9998 + 9999$
$= (1 + 9999) + (2 + 9998) + (3 + 9997) + \ldots$

$+ (4999 + 5001) + 5000$
$= (4999 \times 10\ 000) + 5000$
$= 49,995,000$

(c) −2500.
$1 - 2 + 3 - 4 + 5 - 6 + 7 - 8 + \ldots +$
$4999 - 5000$
$= (1 - 2) + (3 - 4) + (5 - 6) + (7 - 8) + \ldots$

$+ (4999 - 5000)$
$= (-1) + (-1) + (-1) + (-1) + \ldots + (-1)$

[there are 2500 (−1)s]
$= -2500$

2. The digit 0 appears 11 times.
The digit 1 appears 21 times.
Each of the remaining digits appears 20 times.

3. 19 classes.

If there are x students in each class and y classes, then $xy = 361$.

Since $361 = 19^2$, and 19 is a prime number, the only factor other than 1 and 361 is 19. Therefore both x and y equal 19.

Hence there are 19 classes.

4. 60 minutes.

The assumption refers only to seconds in one minute, and not to minutes in one hour.

5. $\frac{91}{216}$.

The chance of not getting a six on one throw is $\frac{5}{6}$. Therefore the chance of not getting a six on

any of three successive throws is
$\frac{5}{6} \times \frac{5}{6} \times \frac{5}{6} = \frac{125}{256}$.

Hence the chance of getting at least one six

$= 1 - \frac{125}{256} = \frac{91}{216}$.

6. 12 minutes.

The cold tap fills the bathtub in 3 minutes.
The hot tap fills the bathtub in 4 minutes.

Therefore, in 12 minutes, the cold tap would fill 4 tubs, the hot tap would fill 3 tubs, and the plug hole would empty 6 tubs in 12 minutes.

Hence, when both taps are running and the plug hole open, one bathtub will be filled in 12 minutes.

Alternatively,

In 3 minutes, the cold tap fills the entire bathtub.
In 1 minute, the cold tap fills $\frac{1}{3}$ of the bathtub.
In 4 minutes, the hot tap fills the entire bathtub.
In 1 minute, the hot tap fills $\frac{1}{4}$ of the bathtub.
In 2 minutes, the entire bathtub is emptied.

In 1 minute, $\frac{1}{2}$ of the bathtub is emptied.

Therefore, in 1 minute, $\frac{1}{3} + \frac{1}{4} - \frac{1}{2} = \frac{1}{12}$ of the bathtub can be filled if both taps are running and the plug hole open.

Hence, when both taps are running and the plug hole open, one bathtub will be filled in 12 minutes.

7. 0%.

The sum of the digits of any of the 9-digit number formed is $1 + 2 + 3 + \ldots + 9 = 45$, which is divisible by 9. Thus the number is always divisible by 9. Hence none of the numbers are prime.

8. 48.

Observe that the greatest number dividing all three given numbers will also divide the difference between any two of the three numbers.

$$308 - 260 = 48$$
$$308 - 212 = 96$$
$$260 - 212 = 48$$

Now, G.C.D. (48, 96, 48) = 48

Observe that $\frac{308}{48} = 6 + 20$; $\frac{260}{48} = 5 + 20$, and $\frac{212}{48} = 4 + 20$

The greatest number is 48, which gives a remainder of 20 in each case.

9. 19 packets with 43 candies in each packet.

If x represents the number of candies in a packet, then x must divide both 301 and 516.

$$301 = 7 \times 43$$
$$516 = 12 \times 43$$

The only number dividing them is 43.
Thus, there must be 43 candies in each packet, with my consuming 7 packets last month and 12 packets this month, giving a total of 19 packets altogether.

10. 22 times.

Midnight	1	The 12th time has to be at noon.
≈ 1.05	2	
≈ 2.10	3	Since we started at midnight, noon is in the second half of the 24-hour day.
...	...	
≈ 9.45	10	
≈ 10.50	11	The minute hand points in the same direction as the hour hand 11 times in the first half of the day. Therefore the total is (11 × 2) = 22 times.
≈ 11.55	12	

Mathematical Quickies & Trickies 20 (p. 124)

1. The man was born in 1180 BC and died in 1163 BC—17 years later.

2. 225 cubes.
The table shows the number of cubes of various sizes

1×1×1	2×2×2	3×3×3	4×4×4	5×5×5	Total
125	64	27	8	1	225

3. The cylinder has the largest volume because the sphere and the cone can each fit into the cylinder.

4. 1:20 p.m.

From 11:12 a.m. to 12:32 p.m., there are 1 h 20 min, or 80 min.

Fraction of the journey representing 80 min

$$= \frac{3}{4} - \frac{1}{3} = \frac{9-4}{12} = \frac{5}{12}$$

$\frac{5}{12}$ of the journey took 80 min.

$\frac{1}{4}$ of the journey would take $\left(80 \times \frac{12}{5} \times \frac{1}{4}\right)$ min

= 48 min.

Therefore Ravin reached Town B at (12:32 + 0.48) = 1:20 p.m.

5. 256 pages.

Pages	Numbers of digits
1–9	9
10–99	90 × 2 = 180
Total	189

Number of pages containing 3 digits

$$= \frac{1}{3}(660 - 189)$$

$$= \frac{1}{3} \times 471 = 157$$

Total number of pages = 157 + 99 = 256

6. 8 minutes.
Hint: LCM(8, 10, 15).

7. 419.
Hint: LCM (2, 3, 4, 5, 6, 7) − 1

8. Note that the 5th square is the large square containing the four smaller squares.

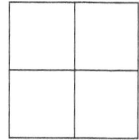

9. Three goats.

Goat	Cow	Chicken
4 legs	4 legs	2 legs

If all 18 heads had only 2 legs each, there would only be 36 legs.

The extra (52 − 36) = 16 legs must come from either goats or cows.

Since every two extra legs come from either goat or cow, we have $\frac{16}{2} = 8$ four-legged animals.

Number of chickens = $18 - 8 = 10$

Number of cows = $\frac{10}{2} = 5$

Number of goats = $18 - 10 - 5 = 3$

Alternatively

Let the number of goats be x.

Then the number of chickens and goats will be respectively $2x$ and $18 - 3x$, respectively.

$$4x + 2(2x) + 4(18 - 3x) = 52$$
$$x = 5$$

Number of goats = $18 - 15 = 3$

10. $2.55.

$1105 = 5 \times 13 \times 17$

Since each bar costs more than 50¢ and less than 80¢ (the selling price), the cost of each bar is $13 \times 5¢ = 65¢$.

Number of chocolate bars = 17

Amount collected from selling chocolate bars
= $17 \times 80¢ = \$13.60$

Profit = $13.60 - $11.05 = $2.55

Mathematical Quickies & Trickies 21 (p. 131)

1. $4 \times (35 + 2) = \$2.80$.

2. $\frac{2010}{2011}$.

Hint:

Consider $\left(1 - \frac{2009}{2010}\right)$ and $\left(1 - \frac{2010}{2011}\right)$.

3. Not 64, but 60 squares.

4. $1 - \frac{1}{2010}$.

Observe that $\frac{1}{1 \times 2} = 1 - \frac{1}{2}$, $\frac{1}{2 \times 3} = \frac{1}{2} - \frac{1}{3}$,

$\cdots, \frac{1}{2009 \times 2010} = \frac{1}{2009} - \frac{1}{2010}$

5. $40.

$20x + 30y = 340$; $y = 5x$
$20x + 30(5x) = 340$
$2x + 15x = 34$
$x = 2$

Amount paid = $\$(2 \times 20) = \$40$

6. (a) 157 whole numbers

Number	Frequency	
102, 112	2	
120 – 129	10	→ Total = 19
132 – 192	7	
200 – 299	100	
302 – 392	19	
402 – 492	19	
Total	157	

(b) 180 times

Number	Frequency	
102, 112	2	
120 – 129	11	→ Subtotal = 20
132 – 192	7	
200 – 210	12	
211 – 220	12	
221 – 230	20	
231 – 240	11	
241 – 250	11	
251 – 260	11	→ Subtotal = 120
261 – 270	11	
271 – 280	11	
281 – 290	11	
291 – 299	10	
302 – 392	20	
402 – 492	20	→ Subtotal = 40
Total		→ 180

7. 36.

Since the number leaves a remainder of 1 when divided by 7, it must be of the form $7m + 1$, where m is some positive whole number.

The possible whole numbers are: 1, 8, 15, 22, 29, 36, 43, 50, ...

Similarly, the number which leaves a remainder of 3 when divided by 11, must also be of the form $11n + 3$, where n is some positive whole number.

The possible whole numbers are: 3, 14, 25, 36, 47, 58, ...

The smallest number common to both sets is 36.

8. 16,000 km.

Four of the five tires were used at any given time.

Thus, each tire traveled a distance of

$$\frac{4}{5} \times 20{,}000 = 16{,}000 \text{ km.}$$

Alternatively,

Since the car traveled 20,000 km, the five tires together drive 80,000 tire-kilometers.

Since each of the five tires is used equally, each tire will go $\frac{80\,000 \text{ tire-kilometers}}{5 \text{ tires}} = 16{,}000$ km.

9. $x = 1$, $y = 1$, $z = 1$.

Since $x = yz$, $y = xz$, and $z = xy$,

we have $z = (yz)(xz) = xyz^2 = z^3$ and so

$z(z^2 - 1) = 0$.

But $z \neq 0$, therefore $z^2 = 1$.

Since z is positive, this gives $z = 1$.

Similarly, $x = 1$ and $y = 1$.

10. There are 2 men, 5 women and 13 children.

$M + W + C = 20$ people; $M + W + C = 20$ pounds of rice.

(Number of men $\times$ 3 pounds) + (number of women $\times 1\frac{1}{2}$ pounds) + (number of children $\times \frac{1}{2}$ pound) $= (2 \times 3) + \left(5 \times 1\frac{1}{2}\right) + \left(13 \times \frac{1}{2}\right)$

$= 20$ pounds of rice.

Mathematical Quickies & Trickies 22 (p. 138)

1. Not 2, but 0.2.

2. $-\frac{49}{16}$.

$-[-(-2) - (-2)^{-2}]^2 = -\left(2 - \frac{1}{4}\right)^2 = -\left(\frac{7}{4}\right)^2$

$= -\frac{49}{16}$

3. 0.

One of the factors is 0.

4. Not $6S = P$, but $6P = S$.

5. 666 lines.

Suppose 4 members were connected, which would need 6 wires to do so.

If there were already 10 members, 10 more wires would be needed to add an 11th member.

Similarly, a 12th member would need 11 more wires, and so on.

So we need $1 + 2 + 3 + \ldots + 35 + 36$, or

$\frac{36 \times 37}{2} = 666$ wires.

6. 10 days.

10 masons can dig 20 holes in 40 days.
10 masons can dig 10 holes in 20 days.
20 masons can dig 10 holes in 10 days.

Alternatively,

10 masons can dig 20 holes in 40 days.
20 masons can dig 20 holes in 20 days.
20 masons can dig 10 holes in 10 days.

7. 1296 rectangles.

For a 2×2 checkerboard there are 9 rectangles:

    4 1×1 squares
    1 2×2 square
    2 1×2 rectangles
    2 2×1 rectangles

For a 3×3 checkerboard there are 36 rectangles

    9 1×1 squares
    4 2×2 squares
    1 3×3 square
    6 1×2 rectangles
    6 2×1 rectangles
    3 1×3 rectangles
    3 3×1 rectangles
    2 2×3 rectangles
    2 3×2 rectangles

A 4×4 checkerboard contains 100 rectangles; a 5×5 checkerboard contains 225 rectangles.

From the sequence (1, 9, 36, 100, 225, ...), the number of rectangles that can be found on an 8×8 checkerboard is

$1^3 + 2^3 + \ldots + 8^3$, or equivalently $(1 + 2 + \cdots + 8)^2$
$= 1296$.

8. 50 applicants.

Since 20 applicants did not pass their English and Mathematics, they were not qualified for the post.

Of the remaining 80 applicants, $80 - 55 = 25$ were not good in English, and $80 - 75 = 5$ were not good in Mathematics.

Therefore, there were only $80 - 25 - 5 = 50$ applicants proficient in both English and Mathematics.

9. 43.

Let n be the least positive whole number.

Then there exist integers x, y, and z such that

$$3x + 1 = n \quad (1)$$
$$5y + 3 = n \quad (2)$$
$$7z + 1 = n \quad (3)$$

From (1) and (3), $3x = 7z = n - 1$.

So there is a whole number p such that

$n - 1 = 21p$ since both 3 and 7 divide $n - 1$.

$n = 21p + 1$

Possible values of n are 1, 22, 43, 64, 85, 106, ...

From the above set, we need to pick the smallest whole number n which gives a remainder of 3 when divided by 5.

This number is clearly 43.

Alternatively,

$n - 1 = 21p$
$n - 1 = (5y + 3) - 1 \qquad$ from (2)
$21p = 5y + 2$
$y = \dfrac{21p - 2}{5} = 4p + \dfrac{p - 2}{5}$

The smallest whole number y is given by $p = 2$.
So $y = 4 \times 2 + 0 = 8$
Therefore, $n = 5y + 3$
$= 5 \times 8 + 3 = 43$

10. 50%.

Mathematical Quickies & Trickies 23 (p. 143)

1. Only three dead birds remained because the others flew away.

2. 3:45:55 p.m. (55 s after 3:45 p.m.).
It takes only 11 cuts to make 12 equal pieces.

3. 96. $[(6 + 6 + 4) \times 6]$
A *palindromic time* occurs between
00 00 h–05 00 h, 10 00 h–15 00 h, and
20 00 h–23 00 h, at a frequency of 6 times per hour.

4. 200 km.

5. Zero dollars.
The coin was counterfeit because the term "BC" could not have been used then. "BC" means "before Christ," and the minter could not know Christ was going to be born 156 years after he minted the coin.

6. 21 cigars.
64 original cigars yield 64 one-quarters (16 new cigars).
These 16 new cigars yield another 16 one-quarters (4 new cigars).
These 4 cigars yield a further one new cigar.
Total number of new cigars = 16 + 4 + 1 = 21

7. The two rectangles are 4×4 and 3×6.
$L \times B = 2 \times (L + B)$
$LB = 2L + 2B$
$LB - 2L = 2B$
$L(B - 2) = 2B$
$L = \dfrac{2B}{B - 2}$

Since L must be a whole number,
$B - 2 = 1$ or $B - 2 = 2$.
$B = 3$ or $B = 4$
$B = 3$, $L = \dfrac{2(3)}{3 - 2} = 6$
$B = 4$, $L = \dfrac{2(4)}{4 - 2} = 4$

8. $5\dfrac{5}{11}$ minutes past 7.
At 7 o'clock, the minute hand is 35 divisions behind the hour hand.

To be on a straight line, the minute hand must gain 5 divisions on the hour hand. Since the minute hand gains 35 divisions in 60 minutes, therefore, the minute hand will gain 5 divisions in $5\dfrac{5}{11}$ minutes.

9. 6 boys.
6 boys move 6 boxes in 6 minutes. 6 boys move 1 box in 1 minute. 6 boys move 60 boxes in 60 minutes.

10. Not 1111, but 11^{11}.
$11^{11} = 285,311,670,611.$

Mathematical Quickies & Trickies 24 (p. 149)

1. 1.75.

2. 4 children.

3. One hour.

If we label the lamb chops 1 to 6, each with sides a and b, then

> 10 min: $1a$ and $2a$
> 20 min: $1b$ and $2b$
> 30 min: $3a$ and $4a$
> 40 min: $3b$ and $4b$
> 50 min: $5a$ and $6a$
> 60 min: $5b$ and $6b$

4. Hardy will win the race.

In the second race, both boys will meet at 10 m from the finish line.
Since Hardy runs faster, he will cover the last 10 m in a shorter time than Littlewood.

5. None.

Because steel is less dense than mercury, all the spheres will float.

6. 44 times.

Every hour, the hour and minute hands form a 90-degree angle twice, except when there are only three between 2 to 4, and 8 to 10.

Thus, in one day, the two hands form a 90-degree $[(2 \times 24) - 4] = 44$ times.

7. Seven.

8. 120 pages.

Pages	Number of digits
1 – 9 10 – 99 100 – 120	9 $2 \times 90 = 180$ 3×21
Total	252

9. $5400.

$$5 \text{ computers} + \text{profit} = \$13,400 \quad (1)$$
$$3 \text{ computers} + \text{profit} = \$8400 \quad (2)$$

$(1) - (2)$: 2 computers $= \$5000$

Thus 3 computers cost $7500.

Replacing the cost of 3 computers into (2), we have
$$\$7500 + \text{profit} = \$8400$$
$$\text{profit} = \$900$$

The only way Robert could sell 22 computers is through 6 deals.
(4 deals of 3 computers + 2 deals of 5 computers)

Therefore, Robert earned $(6 \times \$900) = \$5400$.

10. 27,000,000. (27 million)

Starting with 000 001, 000 002, ..., the numbers 0 to 9 shows up 100,000 times in any of the six positions.

Therefore, the sum for each position is
$(1 + 2 + 3 + 4 + 5 + 6 + 7 + 8 + 9 + 0) \times 100,000$
$= 45 \times 100,000 = 4,500,000$.

Hence, the sum of all digits from 1 to 999,999 is
$(4,500,000 \times 6) = 27,000,000$, or 27 million.

Mathematical Quickies & Trickies 25 (p. 155)

1. January 1, 1901.

The first century A.D. began on January 1, 1. On December 31, 99, ninety-nine years of the first century had elapsed.
To complete the first century, we must add the entire year 100, ending on December 31, 100.
Similarly, the 19th century ended on December, 1900.

2. 78,253.

$10,000,000 = 10^7 = (2 \times 5)^7 = 2^7 \times 5^7$.
Both 2^7 and 5^7 are not divisible by 10.
The two numbers are 2^7 and 5^7.
The sum of the numbers is $(2^7 + 5^7)$, i.e.,
$128 + 78,125 = 78,253$.

3. 1174 years.

No year is numbered zero.

4. 9 cm^2.

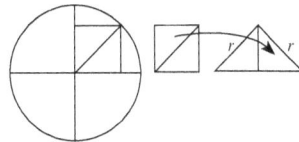

Let r be the radius of the circle.
Then $\pi r^2 = 18\pi$
$r^2 = 18$
Area of square $= \dfrac{1}{2}r^2 = 9 \text{ cm}^2$

5. 1000 km.

1 m = 100 cm = 1000 mm

1 km = 1000 m = (1000 × 1000) mm

1 m³ = (1000 × 1000 × 1000) mm³

One cubic meter cube can be cut into (1000 × 1000 × 1000) pieces of millimeter cubes. When all these smaller cubes are placed on top of each other, the height is (1000 × 1000 × 1000) mm, or $\dfrac{1000 \times 1000 \times 1000}{1000 \times 1000}$ = 1000 km.

6. Bob.

Curtis who fires 3 shots in 3 seconds takes $1\frac{1}{2}$ s between shots, since there are two intervals between the first and third shot.

Similarly, Bob requires 5 seconds for 4 intervals, or $1\frac{1}{4}$ s between shots.

Therefore, Bob will take less time to fire 8 shots. $\left[\text{Compare}\left(7 \times \dfrac{5}{4}\right) = 8\dfrac{3}{4}\text{ s with }\left(7 \times \dfrac{3}{2}\right) = 10\dfrac{1}{2}\text{ s}\right]$

7. $\frac{1}{3}$.

There are four possible combinations from two children:

boy-boy, boy-girl, girl-boy, and girl-girl.

Since the outcome cannot be boy-boy, the chance of two girls is $\frac{1}{3}$.

8. 4-minute intervals.

The trains are evenly spaced along the rail in both directions. Dr. Yan observes the trains at a rate of 30 an hour. Because he is moving towards one stream of trains and away from the other, he sees more trains in one direction than the other (20 to 10), but if he were stationary, he could see 15 an hour traveling each way. The trains, therefore, leave the terminal at 4-minute intervals.

9. 6 members.

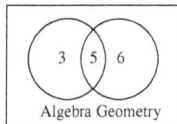

Algebra Geometry

10. 54 minutes.

The wife saved the husband 6 minutes of traveling time each way and thus was picked up at 5:54 p.m., rather than the usual time.

Mathematical Quickies & Trickies 26 (p. 162)

1. 23.

If p = GCD (943, 1357)

then p is a divisor of (943 + 1357) = 2300.

2300 = 23 × 10 × 10
   = 23 × 2 × 5 × 2 × 5

Since neither 2 nor 5 divides 943 and 1357, therefore 23 must be the only divisor of 943 and 1357.

Hence GCD (943, 1357) = 23

2. $\frac{3}{5}$.

3. $x = 2$, $y = 1$.

$$4567x + 5433y = 14\,567$$
$$5433x + 4567y = 15\,433$$

Adding, $\quad$ 10 000x + 10 000y = 30 000

$\qquad\qquad\qquad\qquad x + y = 3 \quad (1)$

Subtracting, 866x − 866y = 866

$\qquad\qquad\qquad\qquad x - y = 1 \quad (2)$

(1) + (2): $\quad 2x = 4$

$\qquad\qquad\quad x = 2$

Substituting $x = 2$ into (3): 2 + y = 3

$\qquad\qquad\qquad\qquad\qquad\qquad y = 1$

Therefore $x = 2$, $y = 1$.

4. 4.

Let x and y be the two numbers.

Then $\dfrac{1}{x} + \dfrac{1}{y} = \dfrac{y + x}{xy} = \dfrac{56}{14} = 4$.

5. $23.

Let us consider the possible bets:

$5 chips can be used to cover 0, 5, 10, 15, 20, 25, 30, 35, …

One $7 + $5 chips cover 7, 12, 17, 22, 27, 32, …

Two $7 + $5 chips cover 14, 19, 24, 29, 34, …

Three $7 + $5 chips cover 21, 26, 31, 36, 41, …

Four $7 + $5 chips cover 28, 33, 38, 43, 48, …

Five $7 + $5 chips cover 35, 40, 45, 50, …

Each combination covers the numbers ending in 0/5, 7/2, 4/9, 1/6, or 3/8.

Observe that the 3/8 combination begins at 28 onwards, so 3, 8, 13, 18, and 23 cannot be obtained.

Thus, 23 is the largest bet that cannot be placed.

6. $9\frac{3}{8}$ min.

In 75 min, the smaller pipe can fill 3 tanks.

In 75 min, the bigger pipe can fill 5 tanks.

In 75 min, both pipes can fill 8 tanks.

Thus, both pipes can fill a tank in $\frac{75}{8} = 9\frac{3}{8}$ min.

7. $3\frac{11}{15}$ days.

The speeds of the couriers are $\frac{4000}{7}$ and $\frac{4000}{8}$.

Therefore, their approach speed was

$4000 \times \left(\frac{1}{7} + \frac{1}{8}\right) = 4000 \times \frac{15}{56}$.

Hence, they will meet in $\frac{56}{15} = 3\frac{11}{15}$ days.

8. 160 km.

In 40 min, the van travels $\frac{60}{60} \times 40 = 40$ km.

Then $80t - 60t = 40$, where t is the time taken.

$t = 2$

Thus, the distance covered is $(2 \times 80) = 160$ km.

9. 36 children.

The lift can hold 120 adults, so 90 adults will use $\frac{90}{120} = \frac{3}{4}$ of its capacity.

This leaves $\frac{1}{4}$ of the lift's capacity for the children.

Since 144 children can fill the lift, the number of children that can be admitted is $\left(\frac{1}{4} \times 144\right) = 36$.

10. 660 m.

There are 23 poles on one side and 22 on the other side.

There are 22 gaps between 23 poles.

Therefore, Math Street is (22×25) m = 550 m long.

Mathematical Quickies & Trickies 27 (p. 169)

1. 200 km.

$t = \frac{x}{v} = \frac{180 + x}{10v}$

$10x = 180 + x$

$x = 20$

The plane goes 200 km, while the ship goes another 20 km.

2. 4.

If $2\frac{1}{2} \equiv 3$, then $10 \equiv 12$.

Hence $\frac{1}{3}$ of 10 = 4.

Alternatively

$\frac{5}{2} : 3 = \frac{10}{3} : x$

$x = 4$

3. 45 trials.

A maximum of 9 trials would be needed for the first key, since if the first 9 fail, the remaining key is the correct key to open the door.

Thus, the worst case would be $9 + 8 + 7 + 6 + 5 + 4 + 3 + 2 + 1 + 0 = 45$.

4. 10%.

Suppose there are 100 political dissidents.

There are $70 + 75 + 80 + 85 = 310$ body parts.

Therefore, at least 10% must have all four body parts.

5. 16 km.

Let's assume the hill to be so tiny that Joe spends a negligible time going up and down it. Then he runs practically the whole time at 8 km/h; so in 2 hours he goes a total of $2 \times 8 = 16$ km/h.

6. 2.

Hint: Let $x = \sqrt{2 + \sqrt{2 + \sqrt{2 + \sqrt{2 + \ldots}}}}$.

Then $x = \sqrt{2 + x}$.

7. 1, 8, and 9.

Let the three numbers be x, y, and z.

Then $xy = 9$ and $yz = 8$.

Since the numbers are digits, they must be whole numbers.

The only possible values for x and y are 1 and 9, or 3 and 3.

The possible values for y and z are 1 and 8, or 2 and 4.

Hence the only possible solution satisfying the problem is 1, 8, and 9.

8. (a) 105; (b) 15.

    (a) The first member shakes hands of the 14 other members.

    The second, who has already shaken hands with the first, shakes hands with the other 13.

    The third, who has shaken hands with the first and second, shakes hands with the 12 others, and so forth.

    Thus, there are $14 + 13 + 12 + 11 + 10 + \cdots + 3 + 2 + 1 = 105$.

    (b) Each of the 15 members hugged with two other members, resulting in $15 \times 2 = 30$ hugs. But, since any member (say, A) hugged with another member (say, B), then B member also hugged with A, thus, we have

$$\frac{30}{2} = 15 \text{ hugs.}$$

9. $\dfrac{400}{3}$ m.

10. 17 kg.

    Hint: Express 1360 into prime factors.

Mathematical Quickies & Trickies 28 (p. 175)

1. *Hint*: There are at least three fairly easy ones, and several others requiring more ingenuity.

    Many answers exist. For example,
$$(1 \times 2 - 3 + 4) \times 5 = 15$$
$$1 - 2 \times 3 + 4 \times 5 = 15$$

2. You should be able to do all except 17, 18, 19, 22, 26, 27, and 29 without using the $\boxed{\sqrt{}}$ button. All but one of these seven numbers can be obtained in the required way.

    Most are easy. $27 = (1 + 2) \times (\sqrt{3})^4$

    22 seems impossible.

3. $(2^{30} - 1)$ cents $\approx 4 \times 10^6$ dollars.

4. 3 broken calculators

    If all 320 sets of calculators were in good condition, the shop owner would pay $320 \times \$0.70 = \$224.00$.

    But he only paid $184.00, so there is a difference of $\$224 - \$184 = \$40$.

    For every broken calculator, he would save $\$7.30 + \$0.70 = \$8$

    So there were $\$40 \div \$8 = 5$ broken calculators.

5. All through the division, we are dividing 9 into numbers whose last digit is 1. After 71, the next higher number ending in 1 is 81, which is itself a multiple of 9; hence we cannot have the figure 8 appearing in the answer.

6. 999999999, 18.

7. 1001, 1,000,001, 1,000,000,001,

    $10,001 \times 99,990,001 = 1,000,000,000,001$

    The number of zeros in the results are 2, 5, 8, 11.

    1,000,000,000,000,001

    $= 100,001 \times 9,999,900,001$

8. 111; 111,111; 111,111,111;

    $3333 \times 33,336,667 = 111,111,111,111$;

    $111,111,111,111,111 = 33,333 \times 3,333,366,667$

9. $2^{83} = 2^3 \times 2^{80}$
$$= 8 \times (2^{10})^8$$
$$\approx 8 \times (10^3)^8$$
$$= 8 \times 10^{24}$$
$$\approx 10^{25}$$

10. 78.

    The numbers around 5 and 2 move forward one position clockwise.

    Therefore, 89 will show up as 78 on the screen.

Mathematical Quickies & Trickies
Bibliography & References

Ang, T. W. (1989). *Mathematically yours*. Singapore: Singapore Science Centre.

Barr, S. (1982). *Mathematical brain benders*. New York: Dover Publications, Inc.

Barr, S. (1965). *A miscellany of puzzles: Mathematical and otherwise*. New York: Thomas Y. Crowel Company.

Birtwistle, C. (1971). *Mathematical puzzles and perplexities: How to make the most of them. London*. George Allen & Unwin Ltd.

Bodycombe, D. J. (1996). *The mammoth book of brainstorming puzzles*. London: Robinson Publishing.

Book, D, L. (1992). *Problems for puzzlebusters*. Washington, DC: Enigmatics Press.

Brandreth, G. (1985). *Everyman's classic puzzles*. London: J. M. Dent & Sons Ltd.

Brecher, E. (1994). *The ultimate book of puzzles, mathematical diversions, and brainteasers*. London: Pan Books.

Chemyak, Y. B. & Rose, R. M. (1996). *The chicken from Minsk*. London: Phoenix.

Devi, S. (1989). *More puzzles to puzzle you*. New Delhi: Orient Paperbacks.

Dudeney, H. E. (1970). *Amusements in mathematics*. New York: Dover Publications, Inc.

Dunn, A. (Ed.). (1980). *Mathematical bafflers*. New York: Dover Publications, Inc.

Edmiston, M. C. (1997). *Fantastic book of math puzzles*. New York: Sterling Publishing Co., Inc.

Gardner, M. (1970). A new collection of short problems and the answers to some of "life's." *Scientific American*, *223*(5), 116-118.

Gardner, M. (1970). The paradox of the nontransitive dice and the elusive principle of indifference. *Scientific American*, *223*(6), 110-114.

Gardner, M. (1971). Quickie problems: Not hard, but look out for the curves. *Scientific American*, *225*(1), 106-109.

Gardner, M. (1971). Ticktacktoe and its complications, and answers to the quickie puzzles. *Scientific American*, *225*(2), 102-105.

Gardner, M. (1986). *Entertaining mathematical puzzles*. New York: Dover Publications, Inc.

Gardner, M. (1966). Puzzles that can be solved by reasoning based on elementary physical principles. *Scientific American*, *215*(2), 96-99.

Jacoby, O. & Benson, W. H. (1996). *Intriguing mathematical problems*. New York: Dover Publications, Inc.

Jargocki, C. P. (1976). *Science brain-twisters, paradoxes, and fallacies*. New York: Charles Scribner's Sons.

Kordemsky, B. A. (1990). *The Moscow puzzles*. Penguin Books.

Krulik, S. & Robert, R. E. (1980). *Problem solving in school mathematics (1980 yearbook)*. Reston, Va: National Council of Teachers of Mathematics.

Longley-Cook, L. H. (1970). *New math puzzle book*. New York: Van Nostrand Reinhold Company.

Maslanka, C. (1987). *The pyrgic puzzler*. London: The Kingswood Press.

Mott-Smith, G. (1954). *Mathematical puzzles for beginners and enthusiasts*. New York: Dover Publications, Inc.

Northrop, E. P. (1980). *Riddles in mathematics*. Penguin Books.

Travers, J. (1941) *A puzzle-mine: Puzzles collected from the works of the late Henry Ernest Dudeney*. London: Thomas Nelson and Sons Ltd.

Trigg, C. W. (1985). *Mathematical quickies*. New York: Dover Publications.

Vecchione, G. (1997). *Math challenges: puzzles, tricks & games*. New York: Sterling Publishing Co., Inc.

Wells, D. (1995). *You are a mathematician*. Penguin Books.

Yan, K. C. (2011). *More mathematical quickies & trickies*. Singapore: MathPlus Publishing.

Yan, K. C. (2009). *Geometrical quickies & trickies*. Singapore: GLM Pte Ltd.

Yan, K. C. (2009). *Get calculator smart*. Singapore: EPB Pan Pacific.

Yan, K. C. (2005). Casting Out Nines. *YG (Young Generation)*, *291*, 6-8.

www.ingramcontent.com/pod-product-compliance
Lightning Source LLC
Chambersburg PA
CBHW080720220326
41520CB00056B/7182